T0189492

Travel Behavior Characteristics Analysis
Technology Based on Mobile Phone Location Data

Fei Yang · Zhenxing Yao

Travel Behavior Characteristics Analysis Technology Based on Mobile Phone Location Data

Methodology and Empirical Research

Fei Yang
Department of Transportation and Logistics
Southwest Jiaotong University
Chengdu, Sichuan, China

Zhenxing Yao
College of Transportation Engineering
Chang'an University
Xi'an, Shaanxi, China

National Natural Science Foundation of China
Department of Social Sciences of Ministry of Education of the People's Republic of China
Ministry of Education of the People's Republic of China
Philosophy and Social Sciences Office of Shaanxi Province
Department of science and technology of Shaanxi Province
Ministry of Science and Technology of the People's Republic of China

ISBN 978-981-16-8010-6   ISBN 978-981-16-8008-3   (eBook)
https://doi.org/10.1007/978-981-16-8008-3

Foreword

Along with the process of urbanization and motorization, the development goals, concerns and countermeasures of transportation are also undergoing great changes. In the future, the field of traffic and transportation engineering will have independent characteristics and integrated interdisciplines, such as engineering science and humanities. Also, the related theories and technologies in this field are at an important development stage in which challenges, explorations and opportunities coexist.

Due to the constraints of resources and the environment, the unprecedented pressure brought by the worldwide rapid urbanization and motorization can no longer be only solved by the construction of transportation infrastructures. With regard to urban transportation, it is also necessary to construct integrated countermeasure systems that can be suitable for many countries' conditions and integrates strategy, policy, planning and construction. Since urban transportation has entered transformation and development stages with equal emphases on construction and management, organically integrating social management and transportation technology system construction and guiding urban transportation modes into the track of sustainable development through rational supply and dynamic regulations have become the challenges that theoretical researchers, managers and engineering technicians have been thinking and exploring painstakingly.

In recent years, in the transportation field, the rapid development of related technical fields has led to many technical concepts undergoing dramatic changes. The integration of information technology and control technology has contributed to the breeding and development of the Internet of vehicle, and the social concern for traffic safety has promoted the application of active and passive safety technologies in both vehicles and traffic technology facilities. "Internet + Service" and "Internet + Traffic" have spawned new transportation service models, such as bike-sharing, time-sharing rental and online car-hailing. Other transportation modes, such as new-type trams and personal rapid transit are gradually coming into our lives. Also, the construction of traffic information systems and the development of the big data theory have created conditions for monitoring and strategically controlling the urban transportation systems with complex adaptation characteristics.

In such a context, traffic engineering researchers, engineering technicians and industry managers all feel that their original knowledge and experience are challenged. They need to re-recognize the problems we face and perform theoretical innovations and technical changes in response to the current technological environment and demand changes.

Facing a rapidly changing and evolving field, this book strives to more quickly reflect relevant theoretical research results and practical experiences, and the following three basic principles were emphasized.

(1) Theoretical Pertinence to Meet the Needs of the Development Stage

World transportation is currently entering a new stage of transformation and development. The current ecological environment and urban living environment concerns require countermeasures with regard to green transportation under the guidance of sustainable development. Under the background of China's urban planning entering the connotative development of total land use control, the contradiction between the growth of traffic demand in built-up areas, which is triggered by land redevelopments and the difficulty of the large-scale expansion of transportation infrastructures, needs to be correctly handled. Moreover, urbanization is entering the stage of urban agglomeration development, which requires the integration of comprehensive transportation systems from the perspective of urban agglomeration and the adjustment of central city transportation modes to adapt to the evolution of functional layouts.

(2) Scientific and Technological Originality in Line with the Development Trend of the International Traffic Theory

The research in traditional fields, such as traffic network flow analysis and traffic behavior analysis, is deepening. The intersection of big data with the complexity theory has given an important impetus to the change in traffic theory. Also, the focus of the urban planning theory from "location space" to "flow space", the relationship transformation of geography, and the study of space behaviors have provided results for reference. The whole disciplinary system is in a trend of change.

(3) Practical Exploration Suitable for the Characteristics of Different Traffic Scenes

Urban transportation strategies, policies and plans are not technical issues, but they form a manifestation of the scientific outlook on development and public policy. Thus, practical exploration under specific management system structures plays an irreplaceable role in theoretical and technical research.

The rapid development of worldwide transportation urgently needs theoretical and technical support, and the practice of worldwide transportation provides the necessary "soil" for the growth of relevant theoretical and technical changes. In the face of the current developments and changes, there is no need to make strict demands to make transportation systems perfect, and the blossoming of all the flowers in research and practice is bound to bring about a colorful future in the fields of transportation science and technology all over the world.

Beijing, China   Guangtao Wang
April 2016

Preface

The era of big data in the world has come. The mining and analysis of massive data will provide valuable information in assisting decision-making and have become hot research frontiers in various industries at home and abroad. Among these, the construction and mining applications of big data environments for transportation are among the most important contents for the development of the big data trend. Location data led by mobile phone positioning are certainly an important core foundation of big traffic data, where a large number of mobile phone users, extensive communication coverage networks and communication event activities generate huge amounts of data, which are very valuable resources. In-depth excavation is expected to infer and reveal the original mechanism and deep-seated characteristics of traffic activities, which will provide an important foundation for analyzing the relationship, characteristics and laws of traffic and urban development in China, which is of far-reaching significance.

Travel demand identification has two major aspects of the basic data of people and vehicles, and the origin is human travel. In the past ten years or so, a large number of good technical means have been accumulated for collecting travel demand data (motor vehicles) at home and abroad, such as by GPS floating cars, loop coils, video detection, radar detection, radio frequency identification (RFID) and other technologies. Also, in many cities, road traffic vehicle information collection and dissemination systems were built. However, the individual "people" factor, which reflects the origin of traffic travel characteristics, has been lacking very effective technical means for tracking travel activity. Traffic flow parameters are only the result of the traffic demand presented on road networks, while individual behavioral characteristics, evolutionary patterns and change patterns are among the determinants directly related to the origin of traffic demand. The shortcomings of traditional residential travel questionnaire data have already caused complaints from traffic engineering colleagues. Such problems include the subjective recall of data errors, organization and implementation difficulties and high costs. At present, despite the research and empirical application of some new technologies performed at home and abroad, there are still more deficiencies, especially with regard to more mature extraction and analysis technologies for individual travel activity data, which can be applied but have not yet been studied. With the rapid development of the 3G/4G-LTE mobile Internet technology, the number of smartphone users has significantly increased, providing carriers with opportunities for tracking and analyzing individual travel activity characteristics. Also, this may bring a huge impact on the development models of traditional transportation industries.

Concerning the analyses of personal travel activities, the application of mobile phone location big data is now mainly based on mobile phone signaling data with regard to performing plenty of application practices. By tracking the information interactions between mobile phone users and base stations in the event of a communication event (calling, messaging, surfing the Internet, etc.), the fuzzy location trajectory of the coverage of base stations is exposed. Then, the activity characteristics of the individual cell phones in cities can be deduced. Finally, the results can be statistically analyzed. Related applications include medium- and macro-level analyses, such as the jobs-housing balance, the distribution characteristics of population mobility, the trip distribution of urban mega traffic analysis zones, typical cross-sectional road traffic flows, which are mostly used for management decision support in urban planning and traffic strategies. However, due to the lack of statistical reliability in the analysis results based on mobile signaling data, there are lots of questions and concerns concerning the accuracy of the data results. At the same time, the accuracy of location data is coarse and cannot provide fine travel chain characteristics, so the quantitative demand for traffic data in traffic modeling cannot be supported.

This book focuses on the high-quality, refined extraction technologies of individual travel data by using individual smartphone sensor data (smartphones have a variety of built-in sensors, and APP can export recorded data, including mobile phone GPS location point coordinates, speed, acceleration, satellite number, accuracy, Wi-Fi access data, etc.). On this basis, we aimed to explore the integration of 3G/4G-LTE new-generation mobile communication network signaling event data (switching, location area update, video, messaging, WeChat QQ, etc.). Such mobile phone data can reflect a mobile phone user's spatial and temporal location changes and movement state change characteristics. With the comprehensive use of wavelet analysis, random forest, cluster analysis and other pattern recognition data mining algorithms, we strived to achieve the refined tracking and extraction of individual travel activity characteristics. The generated report covers a series of detailed "examination reports" of individual travel activities, including travel OD, travel mode, time and location of interchange points, travel time of each travel mode segment and stay time of each residence. It also provides a solid foundation for the optimization and reconstruction of traffic theory models, urban and traffic development plans and management decisions.

Although big data research has received great attention, we need to think deeply about the specific uses and roles of data. We should not just stay in the euphoria of the beautiful "soap bubbles" that big data resources may bring but carry out targeted and in-depth mining to put the data into applications. In view of this, this book also reflects on the application of individual travel activity refinement data and on how to support theoretical models to perform fine calibration and optimization. Some examples include the calibration of traditional four-step traffic model calibration to improve accuracy, transit OD inverse model evaluation and optimization and the new generation of activity-based traffic demand analysis models for empirical calibration. At present, since traditional traffic travel surveys cannot provide the detailed travel chain data required for the calibration of activity-based models, empirical calibration activity models have not been widely used in the international arena.

This book mainly relies on the following researches hosted by the authors: five National Natural Science Foundation of China projects (grant number: 50908195, 52002030, 52072313, 51678505, 51178403), one project supported by the Ministry of Education of the People's Republic of China (grant number: NCET-13-0977), one project supported by the Ministry of Science and Technology of the People's Republic of China (grant number: 2018YFB1600900), one project supported by the Department of Social Sciences of Ministry of Education of the People's Republic of China (grant number: 20XJCZH011), one project supported by the Department of Science and Technology of Shaanxi Province (grant number: 2021JQ-256) and one project supported by the Philosophy and Social Sciences Office of Shaanxi Province (grant number: 2020R035). The book has nine chapters. Chapter 1 introduces the research background, research objectives, application prospects and main contents and summarizes the research features and innovation points. Chapter 2 comprehensively summarizes the recent research and practical cases of mobile phone location data extraction based on mobile signaling data, cellular phone bill data, individual GPS travel trajectory data, mobile Internet check-in data, etc. Chapter 3 introduces the theory of several major data mining algorithms and applications ideas for mobile phone sensor data, including cluster analysis, neural network, support vector machines, Bayesian network, random forest, wavelet analysis, etc. Chapter 4 introduces the APP used in this book to obtain mobile phone sensor data and analyzes the initial data change fluctuation characteristic. Chapter 5 describes the construction method of the "Pedestrian-Traffic Flow-Communication" integrated simulation platform and carries out the evaluation and analysis of the technical effects through simulation experiments. Chapter 6 conducts an empirical study of travel parameter feature extraction technology based on mobile phone sensor data. Also, it uses various data mining algorithms combined with field test data to give a specific process description and accuracy evaluation for the extraction of micro-travel chain feature parameters such as travel mode, travel transition point and travel time. Chapter 7 analyzes the impact factors and sensitivity analysis of the mobile phone sensor data extraction techniques and discusses the effectiveness of the techniques applied under different road traffic conditions. Chapter 8 analyzes the practical application of refined traffic data to improve traffic management and planning, such as improving the four-step model and optimizing bus routes to improve the efficiency of the "last mile" connection, etc. Chapter 9 provides a research outlook, with a brief discussion of important issues for future individual traffic data extraction and the development of big traffic data.

This book was written with the help of scholars and postgraduates, and some important results in the book come from the research and dissertation work of the supervised postgraduates, including the continued research work of Ph.D. student

Zhenxing Yao in the direction of mobile phone big data extraction, and the dissertation results of master's students Da-Kun Zeng, Yu Zhao, Xu Han and Lu Dai, thanking your hard work for providing solid accumulation of results for this book. Ph.D. Yao has continuous work in this research area and has made breakthrough in core analysis technologies such as mobile phone trajectory data mining algorithms based on neural networks, support vector machines, decision trees and other machine learning methods, and traffic pattern recognition algorithms based on modal maxima wavelet analysis. As a result, he was a co-author of the book, with great input in the organization and writing of the entire book. We would also like to thank postgraduates Haihang Jiang, Jianyao Zhou, Haoyi Zheng, Taiyu Lu, Yu Li, Yuan Lin, Lilei Wang and Cong Li for their work, including Jianyao Zhou for organizing Chap. 2, Cong Li for Chap. 4, Haihang Jiang for mainly Chap. 5, Haoyi Zheng for Chap. 8 and Yu Li for Chap. 9.

Due to the limitations of the author's scholarly competence and the fact that big data in transportation is an emerging research direction that is undergoing continuous exploration and development, there are inevitably omissions and deficiencies in this book. Readers are welcomed to criticize and correct them.

Chengdu, China   Fei Yang
February 2022

Contents

List of Figures

List of Tables

Chapter 1
Introduction

1.1 General

Existing individual travel demand survey methods mainly include traditional manual questionnaire surveys, global positioning system (GPS) travel trajectory tracking surveys, and cellular signaling data surveys. These three main approaches have their advantages and disadvantages. In recent years, the main disadvantage of traditional manual questionnaire surveys has been the questionnaire quality. At the same time, most questionnaires rely on the subjective recall of respondents. In recent years, cities in European, American and China, such as Beijing and Shanghai, have actively explored the use of GPS to assist resident travel survey. However, there are also problems such as GPS signal blocking, incomplete tracking data, and inconvenience. A corresponding GPS tracking data mining algorithm has not been developed to effectively extract accurate travel chain information. Travel demand analysis based on mobile signaling data in European and American countries has considerable use in practice. It has gradually become the focus of traffic management departments in large cities in China. However, due to the low location accuracy and the coverage range of the communication base station cell, this method can only detect macroscopic travel demand (such as travel origin—destination (OD) data in the traffic area and workplace and residence distributions) and is deficient in extracting travel chain features at the micro level.

At present, no survey method can be entirely dominant. It is necessary to explore new survey methods for individual travel characteristics. From the status quo and trends, survey data accuracy will become a question that needs to be answered urgently. Individual travel characteristic surveys urgently require more reliable reference data for measurement, judgment, and sampling inspection. This book explores the method of extracting traffic travel characteristics with smartphone sensor data (GPS, accelerometer, and other sensors that record GPS location spatiotemporal trajectory data, acceleration data, satellite numbers, and Wi-Fi hotspots). It also aims to examine the technical process and determine the data rules and algorithm for extracting all individual travel characteristic information, including the travel OD,

© Tongji University Press 2022   1
F. Yang and Z. Yao, *Travel Behavior Characteristics Analysis Technology*
Based on Mobile Phone Location Data, https://doi.org/10.1007/978-981-16-8008-3_1

travel time, travel duration, travel mode, and travel purpose. Furthermore, this book characterizes the accuracy of the extracted data with quantitative empirical research and explores their feasibility as reference verification data for travel feature surveys.

1.1.1 Drawbacks of Individual Travel Survey Methods

Individual traffic travel activity is essential for diagnosing current traffic problems, forecasting traffic demand, and developing transport policies. Existing data surveys and analysis techniques for individual traffic activities are inadequate, especially in terms of data quality precision. Without an accurate data source, no matter how elaborates the traffic model is, it may be useless. At present, it is urgent to supplement and improve the research on the latest technology of refining, objectivating, and performing high—quality, dynamic extraction of individual travel activity data to comprehensively and accurately grasp the source of traffic travel demand.

1.   Traditional Resident Travel Questionnaire Survey

Traditional resident travel surveys are the primary source of current traffic travel data. These mainly include questionnaire surveys, telephone inquiries, and email interviews. With the increasing complexity of event types, deficiencies in accuracy and survey costs have become increasingly prominent, making it more challenging to meet the demands of modern transportation planning and construction in terms of data accuracy and reliability and real—time dynamics.

First, data accuracy is difficult to guarantee against recall bias and error. Traditional survey methods rely on respondents' recollection of the travel process to obtain data, such as the travel mode, travel time, and travel destination. Generally, respondents have a great recall burden. There are often information memory deviations, errors, or unwillingness to cooperate in the implementation, which seriously affect the data quality of resident travel surveys. In addition, the traditional residential travel survey method cannot obtain the refined traffic travel data used to improve the traffic model, such as the departure time, arrival time, and transportation mode transfer time. Data bottlenecks have greatly restricted the development of traffic models and traffic planning schemes.

Second, the cost of the investigation is high, and the organization and implementation are complex. A large—scale travel survey of residents in a city requires the leadership of high—level administrative departments in the government, implementation by multiple city and traffic management departments, and the assistance of subdistrict offices in household investigation. It is challenging to organize and implement the survey. In addition, the survey costs increase due to the need to use many human and material resources. In recent years, the cost of traffic surveys alone in many large cities' comprehensive transportation planning has been as high as millions of yuan.

Finally, the real—time performance of survey data is insufficient. Due to the high cost and challenging implementation of traditional resident travel surveys, a large—scale resident travel survey is usually conducted only once every 5–10 years. Since the founding of the People's Republic of China, Beijing, Shanghai, and other megacities have only been surveyed approximately 4–5 times. A considerable number of second—tier cities may not have carried out a resident travel survey thus far. With the background of dramatic urbanization and traffic demand expansion in China, traditional survey data cannot, in a timely and objective manner, reflect current residents' activity rules. Quite a few traffic demand forecasts use insufficient timely data. Some cities have no relevant data and use only small fragments of data in investigations to estimate traffic planning. The poor objectivity and dynamic quality of traditional traffic survey data directly result in ineffectively matching traffic planning with the actual demand as well as difficulty in truly guaranteeing the implementation effect of planning.

2. Survey of Handheld GPS Terminal Devices

Surveys of handheld GPS terminal devices are a new type of resident travel survey technology that has been explored outside China since 2000. The US and the UK have carried out practical applications that improved the quality of traditional questionnaire survey data and confirmed a good effect. However, most of the studies of the extraction method use single travel transportation data. In China, the prevalence of the multimode transport combination problem makes this method inapplicable. In addition, the cost of GPS equipment is an issue that cannot be ignored. The Government must purchase handheld GPS terminal equipment, and even taking a small sample may lead to high equipment costs. For example, a large city with a population of 5 million, for a travel survey sampling rate of 2%, needs to investigate a population of 100,000; the cost of each GPS device is about 1000 yuan, and the equipment cost can be as high as 100 million yuan. This cost is unimaginable for many urban traffic management departments, which can only be carried out in the foreign city with small population size. However, in China, the scale of the urban population in a city is large. In addition, using handheld GPS devices requires many people to carry out long—term, difficult dynamic travel surveys. Therefore, feasibility can be challenging.

3. Individual Travel Activity Survey Based on Cellular Data

This is a new method adopted by some cities around the world in recent years. It is a kind of aggregate survey technology. For example, some Chinese cities, such as Shanghai, Chongqing, and Tianjin, have used cellular data to track base stations to obtain travel activity characteristics at the macro level, including residence and employment distributions and traffic regional travel OD. This method has played an effective role in the macro—level decision—making management of government departments. However, the cellular data location is only at the "base station village" scale. The scope of a base station's urban radius is a few hundred meters, whereas suburbs are usually 1–2 km in diameter, and location precision cannot support individual travel activity chain analysis. At a deep level, the latest period traffic problems include traffic model calibration, bus line adjustment, interchange demands in the last kilometer of rail transit, and the floating population characteristics of urban traffic. Macro—level data are not sufficient, and traditional resident travel questionnaires are almost impossible to obtain. New technology for fine, objective, high—quality, and dynamic individual travel activity data extraction is needed.

1.1.2 Advantages of Mobile Phone Sensor Survey Methods

With the rapid development of smartphone technology, the number of smartphone users in China has been growing quickly, which provides an excellent opportunity to analyze individual transportation travel activities based on big mobile data. Compared with existing technologies such as traditional resident travel surveys, it has outstanding technical advantages and good application prospects. Individual travel feature extraction methods based on smartphone sensor data are expected to address the shortcomings of existing methods.

1. Improving and Checking the Quality of Manual Surveys in Traditional Residents' Travel

A travel chain's characteristics can be extracted and analyzed using a data mining algorithm on smartphone sensor data rather than by relying only on the respondents' memories. Simultaneously, the two can also be compared and recalled for confirmation, which significantly improves the survey quality. The travel chain data can also be used as reference data to check the quality of a manual questionnaire survey. The sensor data survey method of mobile phone apps is currently in rapid development. At present, the primary purpose is to enhance and improve the existing technological methods to better explore the theory and technology of extracting high—precision and refined traffic travel characteristics. Furthermore, this study provides a more reliable data basis for the theory of traffic demand analysis.

2. Increasing the Comprehensiveness of Mobile Travel Survey Applications

Mobile phone sensor data can compensate for extraction deficiencies based on mobile phone signaling data travel surveys. Personal smartphones are taken as the carrier, which is expected to combine the large sample and macro—level advantages of cellular data with the refinement and micro—level advantages of mobile sensor data, enhance the comprehensiveness of mobile phone surveys of travel characteristics, and contribute to mobile phone surveys.

3. Expanding the Operability of the GPS Travel Survey

On the one hand, the built—in GPS chips in smartphones can avoid the inconvenience caused to respondents by carrying special GPS devices and enhance the operability of GPS tracking travel surveys. Users only need to download a data collection app on their mobile phone and complete data collection by opening and closing the app, with no complex operations required. Meanwhile, mobile phone acceleration, mobile phone service base stations, and other data are added to analyze the individual motion state, which eliminates the effect of indoor GPS signal blocking. Moreover, this makes the application conditions of the technology more elastic.

In summary, the traditional questionnaire survey of travel workers is subject to "sampling error". Cellular data are "fuzzy on a large scale", and they can "fuzzy correctly" explain the lawful characteristics of some groups' aggregation activities at the macro level. Individual smartphone sensor data can be "sampled with precision", parsing detailed information about individual travel activities from all perspectives. If the sample of individuals participating in the survey is large enough (for example, more than 60%), "large—sample correctness" will be achieved; if all urban individuals participate in the survey through a mobile phone app, "all—sample correctness" is expected to be achieved.

1.1.3 Traffic Demand Analysis Model Development Challenges

Current traffic demand analysis technology is mainly based on the "four—step" traffic demand forecasting model established in the 1950s, whose primary function is to meet the large—scale road construction needs of that time. However, the contradiction between resources and the environment is becoming increasingly prominent today. The growth trend of road facilities is subject to many constraints, so it is difficult to effectively solve the traffic problems that constantly appear. Against this background, the idea of solving a traffic problem in many large cities with serious traffic problems has gradually shifted from the construction of new facilities to the comprehensive application of traffic demand management policy. Methods based on individual travel behavior analysis have received increasing attention. In recent years, a large number of studies in this field have begun to appear in China. For example, traffic demand analysis of activities has become a research hotspot. However, when this method is taken from theory to practice, it must have a sufficient amount of accurate individual travel activity characteristic data as support, including the specific departure and arrival times of each trip, the resident time, etc. It is almost impossible to obtain this information accurately through the traditional survey method. The evolution of traffic demand analysis theory needs to extract individual traffic behavior data as the breakthrough point. Through new travel survey technology, traffic behavior can be understood deeply and accurately, which is conducive to developing traffic demand analysis technologies and methods.

1.1.4   New Opportunities in the Era of Traffic Big Data

"Big problems" need "big data". At present, China's urban traffic problem is no longer a simple "small problem" of whether two traffic districts need road connectivity but a "big problem" of upgrading from the point and line to the plane, from the plane to three—dimensional space and from three—dimensional to multidimensional space. An in—depth analysis of these issues depends on the "transportation big data environment" with multiple time, space, and category dimensions. The traffic segment data obtained by the traditional traffic survey method and the information island caused by administrative system segmentation cannot deeply analyze the root of the traffic "big problem," let alone solve the traffic problem by effective regulation means.

Despite the rapid development and application of various traffic data acquisition technologies (coil, video, microwave, floating car, mobile phone positioning, etc.) in the past decade, each method has its advantages and disadvantages in terms of application effect. However, a single technology cannot fully meet the requirements of building an efficient, intelligent transportation system. According to the above characteristics of big data, mobile phone location data will have significant importance in the era of big data. With mobile phone location data as the core, multisource data obtained by traditional and conventional methods such as coils, videos, floating cars, and questionnaires can be integrated to build big traffic data in an absolute sense.

Relying on the accumulated historical data of traffic development can continuously reveal the evolution rules of road traffic flow states and the characteristics, patterns, and rules of residents' travel activity behavior in the spatial and temporal dimensions. From the perspective of travel origin, this paper reveals traffic behavior and its influencing factors, provides empirical evidence for theoretical research on travel behavior, and updates the existing traditional traffic analysis methodology, traffic models, and algorithms. For example, existing traffic quantitative analysis methods are mainly based on modeling, and the large number of model assumptions are bound to reduce their practicability and authenticity. Hence, it is difficult to directly guide the formulation of a large number of traffic management policies. Simultaneously, the traditional method of dividing traffic plots according to the road network is changed. The primary traffic travel analysis unit is the communication base station plot (group) used to determine the purpose of travel OD demand more accurately. Based on this, it is expected that the basic traffic theory will be updated in terms of the characteristics of road traffic flow and driving behavior in China. Then, we derive and propose a set of traffic theories and methods with Chinese characteristics and gradually remove the limitation of directly applying European and American traffic experience theory and methods to a certain extent.

1.2 Target and Values of Mobile Phone Data Based Travel Survey Method

This book is committed to promoting transportation models and theories based on refined traffic travel data and maintaining international frontier hot spots at the academic theoretical level. For a long time, the establishment of transportation theory in China has lacked data support. Although transportation theory has been improved in recent years, it still has low data accuracy and great subjectivity. The refined transportation data studied in this book are expected to compensate for this deficiency, provide the necessary support for the observation of the characteristics of individual travel behavior patterns and the evolution of activity rules, promote the reconstruction and update of the development of transportation theory, and establish a characteristic transportation theory system in line with the reality of China's transportation problems. Meanwhile, many traffic models have been established in China, but the long—term existence of "garbage data" significantly restricts traffic models' application effect. The objective, fine, and dynamic data extraction technology of individual travel studied in this book can improve the four—step traffic model's accuracy by improving data quality and providing data that cannot be obtained by traditional manual questionnaire surveys. It is even possible to obtain a theoretical demonstration of the calibration and verification of the activity—based traffic demand analysis model, which has not been studied well in international transportation. This book aims to improve important aspects of such transportation theories and models and continuously increase China's cutting—edge influence in this research direction.

In terms of application practice, we aim to build a set of new technology and method systems for extracting individual traffic travel activity characteristics based on smartphone app sensor data and integrating signaling event data from the 3th/4th Generation Communication System—Long term Evolution (3G/4G—LTE) next—generation mobile communication network. We develop a smartphone app and analyze the process principles of signaling event data in the next—generation mobile communication network. This paper explores the data mining algorithms of travel endpoint determination, travel mode cutting, and travel mode recognition. We also want to address the fundamental problems and extract objective, refined, and dynamic individual travel activity data. Relying on the research results, we will apply for national invention patents and computer software copyrights to improve the research results' application and promotion level. To meet the new demand for refined traffic travel data in the new era considering the difficulty of specific urban traffic problems in China, this paper will provide an additional decision—making reference for urban traffic planning and management.

The results of this research have good prospects for being replicated and expanded. A process—oriented technology and implementation framework system is established by analyzing travel data extraction technology based on mobile phone sensor data. In practice, traffic survey and analysis technology based on mobile phone sensor data adopts two development modes:

(1)   Application of a residential travel survey. As with the traditional paper ques-
      tionnaire survey, the government can organize it, but this survey technology is
      more convenient and low—cost. The leading unit only needs to put the data
      collection app on the government survey website, and the survey participants
      can download and install it through the website or by scanning WeChat codes.
      During the survey, participants only need to open the app at the specified time,
      close the app after the survey and upload the data to complete the survey.
(2)   Enterprise business applications. For example, when bus companies, subway
      companies, large shopping malls, and other enterprises need to analyze
      passenger flow characteristics, trip OD, shops' hot spots, stay times, etc.,
      they can encourage customers to collect app data by swiping complimentary
      cards on buses and online traffic. Participants are given corresponding rewards
      when they complete a data collection step. This method can also be used for
      long—term data accumulation analysis.

1.3   From Mobile Phone Location Data to Travel Information

Because of problems such as complicated survey methods, distorted survey data, high
survey costs, and the large amount of time required to update data, this book proposes
a new individual travel parameter collection method based on mobile phone sensor
data. First, we aim to develop a mobile app to collect the spatiotemporal trajectory
characteristic data of mobile phone sensors when individuals travel on a combination of multiple modes of transportation. The data quality and data characteristics
under different sampling frequencies, traffic environments, and travel modes are
analyzed and combined in depth to explore efficient data repair and data preprocessing methods. Then, on this basis, the precise framework and processes of new
individual travel chain extraction are determined. The feasibility and effectiveness
of intelligent data mining algorithms such as wavelet analysis, neural networks, and
random forests for travel parameter extraction are explored. Furthermore, through
theoretical analysis, the optimization and innovation of the algorithm are explored.
Finally, a more efficient and accurate travel chain information collection technology
is proposed that is suitable for Chinese cities' transportation modes. This technology
can accurately identify acceptable parameters such as the OD, travel times, transportation mode, transportation mode transfer time, and travel purpose under different
environments, thus effectively improving the quality of resident travel survey data.
Specifically, the main research content is described below.

1.3.1 Mobile Phone Location Data Collection and Analysis

1. Smartphone App Client Development

According to the mobile phone operating system architecture (for example, the Android system), intelligent mobile sensor data acquisition apps are developed. We can connect sensor data acquisition interfaces (such as smart mobile phone GPS locators, accelerometers, and gyroscopes) by programming to read the mobile GPS trajectory (location point coordinates, satellite number, and positioning accuracy), three—axis acceleration, steering angle and so on. When the mobile phone GPS satellite signal is shielded (such as in underground or indoor travel), the app can collect the mobile phone user's current service base station's cell number. At the same time, Wi-Fi access can enable the transmission of the network signal data of individual travel places.

To avoid the impact of automatic shutdown after a long time, app development requires carefull consideration of certain things, such as the data extraction content, collection mode (positioning frequency, whether a third—party interface is applied), data storage format, mobile phone power consumption, and other aspects. Simultaneously, the respondents' influence should be reduced as much as possible to ensure that the mobile phone data collection app has good operability in practice.

2. WeChat Official Account Acquisition Platform

In addition to the research and development of the mobile phone data acquisition app, the data acquisition system of the WeChat platform is developed to facilitate app promotion in the form of an official WeChat account. Users only need to scan the WeChat Quick Response (QR) code to follow the public account to perform data collection. Simultaneously, the privacy protection reputation of the WeChat Company can significantly improve residents' willingness to use the app, which is convenient for use in practice.

3. Construction of the Background Database Management System

The data management system adopts a SaaS framework and standard design to meet users' customized data collection requirements (such as period, content, and frequency customization) and multiuser parallel services. Simultaneously, the Oracle database management system is built to classify and partition the user data in terms of "project name + personal code" and establish a comprehensive index and data buffer mechanism to ensure efficient storage and data reading.

1.3.2 Refined Travel Information Extraction and Collection

Various built—in sensor data in smartphones can represent the spatiotemporal position and motion state change information of mobile phone users. In this book, the mobile phone sensor data recorded in the travel process are derived from the data

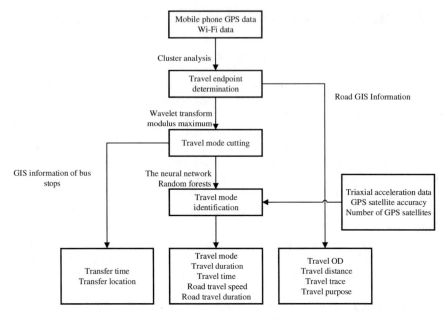

Fig. 1.1 Extraction of the travel route from multimode mobile phone fine sensor data

acquisition system. Integrated mobile phone position change information is reflected by the 3G/4G—LTE next—generation communication network's signaling data. Finally, various data mining algorithms are comprehensively adopted to explore the refined extraction technology and travel data method.

The refined extraction technology of traffic travel features based on mobile phone sensor data is shown in Fig. 1.1, and it includes three main links: travel endpoint determination, travel mode cutting, and travel mode recognition.

1.   Travel Endpoint Determination

With mobile phone GPS data parsed by mobile phone apps and Wi-Fi hot spot data stored in mobile phones, this study intends to comprehensively use a plane clustering algorithm and position point motion state calibration algorithm to determine the travel endpoint. Due to the different GPS location trajectory characteristics of individuals during travel and during stays, there is a linear distribution of track points along the path during travel. There is a certain distance between adjacent continuous position points. The distance between motorized transportation points is more significant, and the distance between nonmotorized transportation points is smaller. However, GPS location points are disturbed near the actual location points due to positioning errors. The distance between adjacent continuous location points will be smaller, resulting in a "clumpy" distribution of the location points on the plane. Therefore, this section uses the plane clustering algorithm to distinguish the "cluster—like" position point group belonging to the travel endpoint according to the distance between the points. At the same time, the Wi-Fi hotspot access location and access time data stored in the smartphone assisted by the mobile phone app analysis, combined with the data of the user's stay time in the base station cell recorded in the mobile communication network, are used as the basis for the residence judgment of the travel endpoint. Finally, the travel endpoint can be acquired.

2. Multiple Travel Modes are Cut

A single trip OD usually contains multiple transportation modes, such as walking + bus/car/rail + walking. There are more diversified combinations of modes of transportation and transfer situations. The result of travel mode segmentation is to separate each transportation mode in a travel trajectory to extract travel features. For example, a trip OD trajectory is divided into walking, bus, and walking segments. Accurate segmentation of the travel modes is essential for refining the extraction of individual travel characteristics. The key is the exploration of the appropriate identification algorithm of transportation mode change points.

The main idea of travel mode cutting is to distinguish the characteristics of velocity fluctuation at position points recorded by mobile phone GPS sensors. Usually, various modes of transportation have different characteristics, such as velocity fluctuation ranges and amplitudes. A change in GPS position velocity is regarded as a fluctuation of the digital signal. The method of finding a "singularity" in wavelet analysis is proposed to identify the traffic mode transition point. As one of the critical research issues in this book, the specific identification algorithm of traffic mode change points will be discussed further in the follow—up.

3. Travel Mode Identification

This book regards the comprehensive characteristics of travel data as different "modes", explores several algorithms in pattern recognition, and classifies different travel modes. This book aims to use neural networks, random forests, Bayesian networks, and other efficient classification methods in pattern recognition for research and to design various multimode combined travel test scenarios, including walking, buses, subways, bicycles, cars, and other significant modes of transportation in Chinese cities. The corresponding sensor data are collected through the developed smartphone app, and the best algorithm of travel mode identification is determined after repeated trial calculation and evaluation. The multimode travel mode recognition algorithm is one of the critical issues in this book, and it will be discussed in the next sections.

After travel mode identification, travel characteristic data can be extracted, including the travel mode in each trip process. Combined with the second link's transformation point results, the departure time, arrival time, and duration of each travel mode can be determined. In addition, with a further combination of travel path information, for individuals using buses, cars, and other motorized travel, the travel velocity and travel time of the corresponding road sections along the journey path can be extracted. In this way, samples are provided for historical road traffic operation characteristics, usually the traffic velocity collection content obtained by GPS floating vehicle collection technology.

1.3.3 'Man-Vehicle-Communication' Simulation Platform Construction and Simulation

The advantage of the simulation platform is that it can determine the effect of the research on the technical method under the environment of various system state parameters, which is conducive to achieving a more profound and comprehensive grasp of the technical method's applicability. This project aims to build an integrated simulation platform that integrates pedestrian simulation, traffic flow simulation, and 3G/4G—LTE next—generation mobile communication simulation. The technical method of refined individual trip data extraction is studied in depth through a multiscene simulation trip test. This test includes four main modules: the traffic simulation module, the mobile sensor data extraction module, the mobile communication network signaling event simulation module, and the technical evaluation and analysis module.

1.   Traffic Simulation Module

According to the configuration of the actual travel data of individuals, individual travel activities are transformed into multimode travel simulation trajectory data with existing traffic simulation models and software. Accurate individual travel data are configured, including the travel OD, travel mode, travel path, travel time, and stay time. Walking travel trajectory data (travel location coordinates, acceleration, etc.) are obtained through pedestrian simulation. The simulation results can reflect the pedestrians' interactions and the interactions between pedestrians and surrounding obstacles and then determine pedestrians' position movement by looking for the objective cost function with the minimum perceptible value. The traffic simulation software's corresponding vehicle travel trajectory data confirm the flow motion parameters' characteristics when using the vehicle transportation mode to travel.

2.   Mobile App Sensor Data Extraction Module

Due to GPS positioning errors in practice, the complete trajectory data of individual travel generated by the traffic simulation are processed by position point coordinate perturbation. Then, the trajectory data of individual multimode travel simulations are obtained. Furthermore, according to the extraction method of refined individual data, the trajectory data of individual multimode travel simulations generated in the traffic simulation module are processed according to the three critical links of travel endpoint determination, travel mode cutting, and travel mode identification. The individual travel OD, travel mode, travel time, and other characteristic data are obtained to compare and evaluate actual travel data and mobile communication network signaling data.

3.   Mobile Communication Network Signaling Event Simulation Module

This module aims to generate the base station cell's location simulation data, including mobile signaling events during individual travel. The source of travel survey data based on mobile signaling data is obtained by mining and analyzing the simulation data. The simulated signaling data can be compared with the results of trip feature extraction from mobile app sensor data, and we explore how to integrate it to improve the trip feature data accuracy. Finally, we can select a particular urban area, build the network according to the 3G/4G—LTE next—generation mobile communication network's layout planning principles, and configure the corresponding communication network parameters.

4.   Technical Assessment Analysis Module

Based on the above three modules, various influencing factors can be set, such as the travel track positioning frequency, road traffic status, combination of various modes of transportation, and frequency of mobile communication activities, to extract the behavioral characteristics of travel survey sources based on the mobile app. Simultaneously, the individual behavior characteristics based on analyzing 3G/4G—LTE mobile communication network signaling events are compared with actual travel data. Additionally, to analyze the advanced extraction methods' precision and sensitivity under various influencing factors and explore the corresponding travel features, the individual behavior characteristics based on 3G/4G—LTE mobile communication network signaling event analysis are compared with actual travel data. Finally, mobile phone sensor data and mobile communication network data are integrated to explore ways to improve individual travel survey accuracy.

1.3.4   Empirical Study and Performance Evaluation

This book explores the application of delicate traffic survey data in traffic theory optimization, traffic planning practice, and management. We explore the value of delicate travel data, quantitatively clarify the role of data, and ensure data collection effectiveness. In this process, the data requirements of theory and practice are fed back to the data collection stage, making it convenient to put forward new data collection requirements to better meet the needs of theoretical development and improve the practical application effect. Acceptable individual trip data are mainly applied to support the subdivision of traffic districts; four—stage model parameter calibration, bus OD assessment, bus network optimization, tracking of passenger flow sources and direction analysis, and activity—based traffic model construction are carried out.

1.3.5   Challenges Faced by Travel Information Detection and Analysis

According to the above research content, the fundamental scientific problems to be solved by this project mainly include the following four areas:

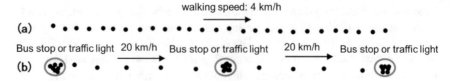

Fig. 1.2 Walking and bus or car travel mode position coordinate point trajectory density

1. Trip Endpoint Identification Algorithm Combining Plane Space Cluster Analysis and Location Point Resident State Recognition

This research aims to use a two—plane space cluster algorithm and location point resident state recognition to solve the fundamental problem of travel endpoint determination.

(1) Plane Space Cluster Analysis

To a certain extent, a mobile phone GPS location trajectory reflects an individual travel activity pattern, and parameters such as the density and distance of the registration point can be better applied to individual travel state recognition. The application of the density clustering algorithm in plane space to identify stopovers is as follows: first, it is determined whether the distance and the user's residence time meet the threshold within a particular area. Then, the difference in the densities of the anchor points is calculated. Finally, the high—density area trajectory (the stopping point) and low—density area (the moving point) are segmented, as shown in Fig. 1.2.

(2) Location Point Resident State Recognition

The resident state of position points can be identified by the plane distance variation characteristics of two adjacent GPS coordinate positions in continuous time. In the same actual position, the positions recorded by the GPS at different times vary due to positioning errors. Additionally, the distance between points at adjacent positions may be relatively large in the process of travel. For example, the motorized travel mode has a more significant point distance than the nonmotorized travel mode, as shown in Fig. 1.2. d is the spatial distance of adjacent coordinate positions in a continuous time series, and α is the critical state characteristic threshold (determined by statistical analysis and comprehensive analysis based on measured GPS data in combination with the positioning accuracy). When $d < a$, the position point is determined as a stay state.

2. Key Calibration Algorithm of Traffic Transfer Points Based on the Wavelet Transform Modulus Maximum Algorithm

This is one of the critical issues and innovative highlights of this book. From the international research frontier perspective, a practical transfer point determination algorithm has not been found in the past ten years. The wavelet transform modulus maximum algorithm has an outstanding ability to recognize unstructured data and perform signal singularity detection. Since the mobile phone app sensor's travel velocity characteristics are similar to the signal fluctuation characteristics, the transfer point's data feature mutation during the travel process is also very similar to the signal wave "singularity" characteristics. Therefore, this study aims to explore the application of the wavelet transform modulus maximum algorithm in transportation transfer point identification, the breakthrough of singularity detection, and the key steps of the optimization of significant noise reduction.

3.   Travel Mode Recognition Algorithm Based on the Pattern Classification Method

We mainly explore the related algorithms in pattern classification to perform multi-mode traffic identification, such as neural networks, random forests, and support vector machines. Because the accuracy of traffic mode identification involves the efficiency of the algorithm itself, the frequency of data collection, the choice of traffic mode calculation attributes, the level of road section congestion, the quality of the data, and so on, the difficulty of this stage is to comprehensively explore the traffic mode data characteristics and data processing steps under different conditions and propose an optimal traffic mode identification method suitable for different situations. Additionally, due to the loss of GPS data such as underground travel, it will need to be combined with the change data of the cell base station in the mobile communication network for judgment, as shown in Fig. 1.3.

4.   Calculation Theory of the Signal Coverage Boundary of a Mobile Communication Cell and Its Correlation with the Spatial Mapping of Traffic Cells

Mobile changes in massive mobile location data at the cell level of communication base stations can reflect the characteristics of residents' travel activities. However, due to the inconsistency between communication and traditional traffic, such potential data cannot be used directly for modeling and analysis. Due to the fluctuation of the base station signal intensity and the overlapping of multiple base station cell signals covering the same traffic cell, it is difficult to establish the mapping relationship with the traffic cell. There is no transparent theoretical system for calculating the base station cell boundary in the field of mobile communication. Therefore, based on the theory of signal strength propagation and attenuation in communication, mobile simulation and signal measurement technology should calculate the base station cell's

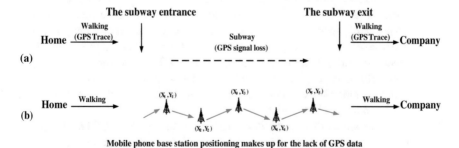

Mobile phone base station positioning makes up for the lack of GPS data

Fig. 1.3 Travel in metro phone GPS data sequences complementary to the base station changes the cell travel mode determination

stability probability boundary range. The spatial mapping correlation between the base station cell's coverage and the traffic cell is established by integrating the cross—disciplinary features. Solving this crucial problem is an essential basis for effectively using mobile location big data to construct traffic models for travel demand analysis.

1.4 Summary

Led by mobile phone sensor data and assisted by mobile communication network signaling event data, this book focuses on the refined extraction of individual traffic travel features, their optimization, and the reconstruction of theoretical models. The technical route adopted is shown in Fig. 1.4. It mainly includes smartphone app sensor data collection, signaling data analysis of 3G/4G—LTE next—generation mobile communication, a refined mining algorithm of individual travel characteristics, traffic model optimization, and other content.

The specific technical route steps and corresponding technical means are as follows:

(1) The research positioning is determined based on sorting the literature review research of the subject, and the main ideas, research content, and critical issues of the subject are clarified.

(2) A smartphone sensor data acquisition app is developed and debugged, and it can export the GPS location point trajectory, velocity, triaxial acceleration, number of satellites, positioning accuracy, Wi-Fi data, and other data. These data can be used as the primary source of refined extraction and mining of individual travel characteristics in the follow—up.

(3) The signaling data of the 3G/4G—LTE next—generation mobile communication network are analyzed, including the data of soft handover, location update, Short Message Service (SMS), WeChat and QQ, and other online events, and fusion assists in subsequent individual trip data mining and the construction of an individual trip integrated simulation platform.

(4) Based on mobile phone data sources, various data mining algorithms in the field of pattern recognition are comprehensively adopted to explore refined feature extraction and mining methods for individual travel. These mainly involve two areas: experimental plane space cluster analysis and a resident point state determination algorithm. The wavelet transform modulus maximum algorithm is used to calibrate the transfer point of the traffic mode and perform travel mode cutting. We explore random forests, neural networks, and other methods of transportation travel mode classification to achieve the refined extraction of individual travel characteristics.

(5) Pedestrian simulation, traffic flow simulation, and 3G/4G—LTE next—generation mobile communication network simulation are integrated to build an individual travel integrated simulation platform and generate pedestrian simulation trajectory data and vehicle simulation trajectory data. This platform

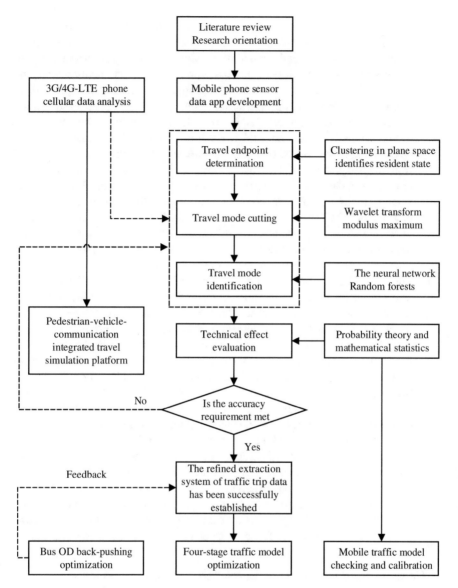

Fig. 1.4 Overall technology roadmap

generates pedestrian communication signaling data based on MATLAB software and the processing mechanism of mobile phone communication events in a 3G/4G—LTE environment. Simultaneously, combined with the statistical probability method, the simulation platform can assist in the simulation precision evaluation of the refined travel data extracted in the previous step. Suppose the precision effect reaches the expectation, which significantly improves the existing travel survey method and meets the refined data demand of traffic planning practice and the theoretical model. This indicates that the refined extraction technology and method of individual traffic trip features have been successfully established. If the accuracy does not meet the requirements, the method needs to return to the data mining algorithm section for adjustment.

(6) The improvement and optimization of traditional traffic models based on the essential traffic demand reflected by the refined data of the travel characteristics are studied. On the one hand, in the four—step traffic demand model, the possible improvement and optimization effects of the refined data are explored from the perspectives of refining traffic plots, improving the quality of characteristic travel data, travel routes and traffic allocation, etc. Meanwhile, given the strategic focus of public transportation development, this paper puts forward the problem of bus OD back—pushing model evaluation, the fusion and correction of accurate bus OD small—sample data and incomplete bus IC card—swiping large—sample data, and the improving effect of the bus OD on bus passenger flow prediction and bus line network optimization.

(7) The book focuses on applying individual refined travel data in activity—based traffic demand analysis models and breaking through the long—standing obstacle of not calibrating the activity model empirically. Simultaneously, the traditional four—stage traffic demand forecasting model is promoted to a new generation of activity—based traffic demand analysis models. A probabilistic model of mobile phone calling behavior is constructed. The individual travel sequence pattern is deduced, and the checking standard based on the mobile traffic demand analysis model is studied and established. These issues have not been thoroughly examined in current international research, which is also one of the research emphases and characteristic innovation points.

(8) The book focuses on breaking through the existing limitations of separating data extraction from the theoretical model application and increasing the process of feeding back data requirements in theoretical model application to perform refined extraction of individual travel characteristics. Simultaneously, for the newly developed data application requirements in the research process, we explore whether it is possible to obtain the data in the data extraction stage and establish an interactive feedback mechanism to ensure the research results' effectiveness.

The features and innovations of the research in this book are as follows:

(1) This study analyzes the signaling event parameters of the next—generation communication network, constructs a 3G/4G—LTE mobile communication network environment, and develops app software for recording sensor data. Mobile phone sensor data, as the leading fusion of mobile communication network signaling data, are proposed to build a traffic big data environment with mobile phone location data as the core, and individual travel activity characteristics and the refined extraction of new technologies and methods are researched.

(2)   Coupling three models with entirely different theoretical principles of pedestrian simulation, traffic flow simulation, and mobile communication simulation, a "pedestrian—vehicle—communication" integrated simulation platform is innovatively constructed to generate individual travel activity tracks. By breaking through the boundaries of various fields, research methods are innovated, and interdisciplinary methods are applied more creatively.

(3)   By comprehensively exploring the applicability of various data mining methods in individual travel mode identification, such as the wavelet transform modulus maximum algorithm, plane clustering, neural networks, random forests, and support vector machines, classical methods are innovatively applied to new problems. The extensive traffic data analysis concept will be gradually implemented in concrete practice, not just staying on paper.

(4)   The research features are to build an integrated interactive feedback research model of individual travel data mining extraction and theoretical model application, effectively compensating for the lack of independent data mining and application research. This model can fully evaluate the practical and theoretical value of travel data and guide data mining according to feedback regarding application requirements.

Chapter 2
Literature Review

2.1 Types and Characteristics of Mobile Data Based Travel Survey Method

Using mobile phone location data to analyze residents' travel characteristics can enable effective investigation of the technical defects of traditional resident travel questionnaires with high data accuracy, lower survey costs, and better development prospects. According to the data source, unique travel characteristic analysis technology based on mobile phone location data can be divided into mobile phone sensor data analysis technology, mobile phone signaling data analysis technology, and mobile phone Internet check-in data analysis technology.

2.1.1 Mobile Phone Sensor Data Based Travel Survey Method

Mobile phone sensor data analysis technology mainly utilizes GPS, accelerometers, gyroscopes, and Wi-Fi smartphones such as built-in sensor chips. Through the development of unique data sampling apps, individual travel information, including cell phone GPS data (travel time, coordinates, travel velocity, etc.), three-axis acceleration data, the number of gyroscope steering angles, and Wi-Fi connection information, is acquired. On this basis, a variety of data processing and mining algorithms are integrated and applied to extract accurate individual traffic trip characteristic information.

The application of mobile phone sensor data in traffic travel analysis is still in the exploration and development stage. In recent years, with the development of smartphone positioning performance and data mining methods, research in this field has become a direction for many institutions and scholars around the world. At present, the application of mobile sensor data analysis technology in the field of transportation mainly focuses on the extraction of individual traffic travel chain information,

© Tongji University Press 2022
F. Yang and Z. Yao, *Travel Behavior Characteristics Analysis Technology Based on Mobile Phone Location Data*, https://doi.org/10.1007/978-981-16-8008-3_2

including travel time, travel distance, travel OD (travel start and end points), and travel mode. The accuracy of data extraction is significantly improved compared with the traditional questionnaire survey method. This technology can compensate well for the low positioning accuracy of other types of mobile phone positioning data and can better support the extraction of refined traffic parameters. At present, mobile phone sensor positioning accuracy is generally up to 10–20 m, and some high-performance GPS chips can even reach 3–5 m. Due to the urban population density and the complexity of China's traffic environment, it is necessary to identify multiple combined travel modes and travel purposes under compound land use and support urban traffic planning and management with refined traffic data. Therefore, combined with the travel characteristics of urban residents in China, there are good application prospects for developing a complete set of refined traffic parameter extraction and analysis technologies suitable for diversified traffic travel environments.

2.1.2   Mobile Phone Signaling Data Based Travel Survey Method

In communication, when a user calls or uses SMS or other communication services, the communication network management center can directly or indirectly record the mobile phone user's location change information. At present, mobile phone signaling data are mainly obtained from the data management centers of mobile communication operators, including the mobile phone user number, communication event type, timestamp, base station number, location area number, and other information. The formatted content is shown in Table 2.1.

Table 2.1 Cellular data content and format

Number	Name	Length	Description	Type
1	MSID	32	The user's unique ID	Uniquely identified cell phone (Temporary Mobile Subscriber Identity (TMSI))
2	TIMESTAMP	14	Synchronized with GPS time	YYYYMMDDHHMMSS For example: 20,051,125,112,203
3	LAC	5	Location area number	–
4	CELLID	5	Cell number	–
5	EVENTID	3	Event type	Includes call, send SMS, receive SMS, hang up, etc
6	AREA_CODE	4	Where the phone belongs	Use the area code to identify where your phone belongs
7	TOLL_TYPE	1	Long-distance type	–
8	ROAM_TYPE	1	Roaming type	–

Traffic survey technology based on mobile signaling data can compensate for the lack of traditional questionnaire survey data to a certain extent and has the advantages of low survey costs, large sample size, and comprehensive coverage. Its data accuracy is related to the density of base stations and the coverage range of base stations. In general, the positioning accuracy in the urban interior is 50–200 m and that in suburban areas is 1000–2000 m, so the positioning accuracy of mobile phone signaling data is lower than that of mobile phone sensor data.

At present, mobile signaling data are mainly used to analyze macro—traffic travel characteristics and travel rules. Research and practice around the world have proven that mobile signaling data have a good application effect for the spatial distribution of the population, the distribution of residents' employment and residences, the analysis of intraregional passenger flow, and the interregional passenger exchange capacity. With the advent of the 3G/4G communication era, the number of individual mobile phone users continues to rise. There are still good application prospects for analyzing the characteristics of urban residents' travel activities using mobile phone signaling data.

2.1.3   Mobile Phone Social Network Data Based Travel Survey Method

With the improvement of mobile phones' wireless communication network velocity and the rapid popularization of smartphones, social networking software, and e-commerce software, the location-based services network (LBSN) for individual traffic travel behavior analysis technology has gradually become a popular frontier.

Mobile Internet check-in data are data with locations generated by users' "check-in" behavior on an LBSN through mobile devices, and they contain multidimensional information such as time, location, and social interactions. Combined with geographic information system (GIS) technology, through visualization, fusion clustering, regression analysis, and gravity modeling, the data can yield the user travel movement trajectory, spatial range, location distribution law, travel purpose, activity type, and other information characteristics. The data mining algorithm can also extract the spatial and temporal trajectory, travel OD, regional population density, job-housing balance, etc.

At present, most of the research related to mobile phone check-in data is in the exploratory stage, and there is still ample space for development in deep data mining, data fusion analysis, and other areas.

2.2 Overview and Summary of Existing Researches and Applications

For a long time, the traditional paper questionnaire survey is a widely used method of individual travel information survey. However, this method has many defects such as low data quality, high survey cost, difficult organization, and long update cycle. Since the 1990s, experts and scholars have been working on extracting and mining travel information and characteristics of residents using mobile signaling data. Due to the large number of mobile phone users, high penetration, low investment cost and real-time dynamic data, the analysis technology of medium and macro traffic characteristics and rules based on mobile phone signaling data has rapidly developed.

According to the research and practical experience in recent years, the accuracy of mobile signaling data is limited by the base station density and base station coverage, which is more suitable for medium and macro traffic travel law and magnitude feature recognition. For the fine travel chain information extraction required by the traffic fine design and traffic model optimization, the data accuracy is insufficient.

Individual travel information extraction technology based on mobile phone sensor data has received increasing attention by scholars at home and abroad due to its higher positioning accuracy, lower cost and lighter burden compared to base station positioning of mobile phone signaling. Meanwhile, with the popularity of smart phones and social software, the analysis technology of individual travel characteristics based on the location check-in data of mobile social networks has gradually attracted attention. The individual travel information extraction technology based on multi-source data fusion is also a research hotspot in the field of traffic survey.

The research on the development of traffic analysis technology based on mobile location data is summarized in Tables 2.2, 2.3, 2.4 and 2.5.

2.3 Individual Travel Behavior Analysis Based on Mobile Phone Signaling Data

2.3.1 Dynamic Monitoring of Residents' Travel Activities

In 2005, MIT and Mobilkom, an Austrian mobile company, jointly launched a real-time project in Graz [1]. The project obtained the number of mobile phone users and geographical distribution information through a perfect communication network system. The preliminary results of the project show that the mapping of mobile data can open up an unprecedented perspective of the city to obtain a new city model: real-time city. Second, based on the mobile characteristics of users recorded by mobile phones, this method can provide new inspiration about the relationship between the information and communication technology and contemporary cities.

Table 2.2 Analysis of individual activity patterns based on cell phone signaling data

Data type	Research content	Time	Place	Technical mode	Main achievements
Analysis of individual activity rules based on mobile signaling data	Dynamic monitoring of residents' activities	2005	Austria Graz	GIS map matching	By using mobile positioning data, it provides a platform that can reflect the dynamic changes of cities and can rebuild and expand their natural features
		2006	Italy Rome Milan	Clustering fusion	This paper explores the internal relationship between the service intensity of communication base stations and the density of regional residents. The research shows that there is a positive correlation between the intensity of individual activities and the service intensity of communication base stations in a region
		2008	USA	Theory of biological behavior; Mathematical statistics	According to the spatiotemporal positioning data, the probability distribution of travel distances is fitted, and it is found that residents show strong regularity in travel distances and travel destinations. The results show that the clustering of individual mobile location data can provide empirical data for the basic theory of travel activity behavior
		2010	Estonia	GIS map matching	Taking Tallinn, the capital of Estonia, as an example, this paper studies the commuting behavior of suburban residents and the tourism behavior of tourists through the measured mobile phone positioning data. The results show that mobile location data can enable a better understanding of the relationship between urban space, travel behavior and related factors, and can enable better configurations of public service facilities
		2014	Singapore	Rule-based pattern partition	This paper explores the feasibility of using mobile phone positioning data to analyze the daily travel activities of large-scale urban residents and compares the pedestrian travel mode and motor vehicle travel mode
		2014	China: Beijing	GIS map matching; Spatiotemporal clustering	Based on mobile positioning information, this paper analyzes the temporal and spatial distribution of traffic demand and analyzes the characteristics of the job and residence distributions of rail users and the temporal and spatial characteristics of population flow

(continued)

Table 2.2 (continued)

Data type	Research content	Time	Place	Technical mode	Main achievements
		2014	China: Wuxi	GIS map matching cluster analysis	Based on a travel survey of mobile phone signaling data, this paper analyzes the long-term historical travel trajectory and spatial activity range of residents and obtains the distribution law of occupation and residence in the core area of Wuxi City
		2015	China: Shanghai	Time threshold method; Information entropy method; Discrimination of the relative value of time	Based on the signaling data of mobile phones, three methods are used to identify the residence of residents according to their different activities in the daytime and at night. The recognition results show that the effect of the information entropy method is better and the recognition rate is higher
		2016	China: Shanghai	GIS map matching cluster analysis	The distribution of a population, occupations and residences is analyzed by using mobile communication data. The characteristics of commuter and tidal traffic and the characteristics of rail transit passenger transfer are analyzed
		2016	China: Chongqing	GIS map matching cluster analysis	Based on the signaling data of mobile phones in the main urban area of Chongqing, this paper evaluates the balance of work and housing among groups. The travel situation of residents in different regions and different time periods is studied
	Analysis of regional and cross-sectional passenger flow	2005	China: Beijing	Cluster analysis	Based on the time distribution characteristics of mobile phone traffic, urban activities and land use are analyzed and divided to perform a quantitative transformation from traffic to the active population
		2009	China: Shenzhen	Road network visualization based on space	According to the location change of mobile phone users in the range of a base station, it is mapped to the traffic area to obtain the corresponding OD data. This can provide support for urban transportation, construction, planning and management

(continued)

Table 2.2 (continued)

Data type	Research content	Time	Place	Technical mode	Main achievements
		2013	China: Tianjin	GIS map matching cluster analysis	Based on the mobile phone data provided by operators in Tianjin, the travel information of residents is extracted, and the distribution of work and residences, the travel characteristics of commuters and the OD distribution of interregional travel are analyzed, which fully demonstrates the effectiveness and practicability of mobile phone data as a traffic survey method
		2013	Belgium	Spatiotemporal clustering	Based on continuous mobile tracking data, this paper analyzes the internal relationship between the individual travel purpose and traffic behavior decision-making. The accuracy of individual travel activity prediction based on mobile location data can reach approximately 70%
		2014	China: Wuxi	GIS map matching cluster analysis	According to the signaling data of mobile phone users in Wuxi City, the OD and passenger flow corridor of residents in each area of the city are identified, and the characteristics of passenger flow in Wuxi City are obtained
		2014	Bangladesh: Dhaka	GIS map matching cluster analysis	CDR is used to generate a complex OD matrix, and it is combined with a micro-scale traffic simulation platform to determine the best matching proportion factor from traffic survey data
		2015	China: Shanghai	GIS map matching visualization	Taking the inner ring road as the research section, the OD of the section in and out of the morning and evening peaks is obtained, and the feasibility and applicability of using mobile phone positioning data to obtain the traffic OD are determined; this can provide a reference for traffic planning and management
		2016	China: Chongqing	GIS map matching cluster analysis	The OD matrix of commuting trips obtained from mobile signaling data is used to analyze the trip distribution characteristics among groups, which reflects the development characteristics of urban layout and the accuracy of the OD

Table 2.3 Analysis of individual travel behavior characteristics based on cell phone sensor data

Data type	Research content	Year	Place	Technical mode	Main achievements
Analysis of individual travel behavior characteristics based on mobile phone sensor app data	Traffic mode identification	2004	Denmark	GPS data mining and analysis technology	Using a handheld GPS positioning instrument combined with an electronic questionnaire, this paper explores the temporal and spatial movement characteristics of individual travel and pedestrian travel characteristics. It is found that the GPS positioning point trajectory is not accurate due to building occlusion, and the density of GPS positioning points is different in different traffic modes
		2006	Canada: Toronto	Pattern recognition algorithm based on fuzzy logic, GPS + GIS map matching	The GPS + GIS integrated system improves the accuracy of multiple travel mode identification. Although the data processing time of the integrated system is long, it highlights the advantages of an interactive analysis system
		2008	Netherlands	Data collection method using web information interaction	The method based on a GPS can record more data, and the interactive confirmation of the website can greatly improve the accuracy of data identification
		2010	USA	Data acquisition based on a mobile GPS positioning chip	Data acquisition software based on a mobile GPS positioning chip is developed. The average recognition rate of the traffic mode is 82.14%; the highest is walking, 95%, and that of cars is 91.25%. The carbon emissions of various traffic modes are evaluated
		2012	USA: Cleveland, Ohio	Mining analysis technology based on GPS	By filling in the online travel log and checking the travel purpose and mode identification results, this paper explores the feasibility of GPS traffic survey technology completely replacing the traditional questionnaire survey
		2013	England: London	Bayesian network	The accuracy of traffic mode identification is greater than 90%, which proves that the acceleration data can greatly improve the accuracy of traffic mode identification. The disadvantage is that considering only a single travel mode is not sufficient for studying a combination of multiple travel modes

(continued)

Table 2.3 (continued)

Data type	Research content	Year	Place	Technical mode	Main achievements
		2015	Iran: Teheran	Neural network, decision tree, random forest, etc	The algorithm mechanism, sample size, identification parameters and model assumptions have the most significant influence on the accuracy of traffic mode identification; The results show that the recognition accuracy of the random forest model for bus, car and walking travel modes is the highest, reaching 87.93%, 97.68% and 90.33%, respectively. When the velocities are similar, the recognition rate is not high
		2015	Netherlands	Mobile smartphone app	Automatic detection of the departure and arrival time, origin and destination, mode of transportation and travel purpose; this reduces the burden of many interviewee records, but 20–25% of the trips are not detected, resulting in high power consumption
	Travel OD	2007	USA: Lexington	Based on the time threshold, distance threshold, direction and road network matching	The average recognition accuracy is close to 91%
		2013	Switzerland	Probabilistic map matching algorithm	This paper explains the limitations of the traditional GPS probabilistic map matching algorithm and puts forward the idea of extracting residents' travel chains by fusing mobile phone Bluetooth, wireless base station, gyroscope, magnetometer and other types of data inside the mobile phone
		2013	China: Beijing	Individual travel state recognition method based on a threshold rule	Travel endpoint recognition: with the increase of time reading and distance reading, the recall rate of stop recognition decreases and the precision rate increases; this solves the problem of large-range (1–2 km) positioning drift and jitter in mobile phone positioning data

(continued)

Table 2.3 (continued)

Data type	Research content	Year	Place	Technical mode	Main achievements
		2015	Japan: Nagoya	C- Density-Based Spatial Clustering (C-DBSCAN) algorithm (combination of clustering and a support vector machine)	The accuracy of stopping point identification can reach 90% in the first step, and in the second step, the accuracy of stopping point identification can reach 96% by using a support vector machine, which cannot fully consider non-purposeful stopping caused by vehicle accidents
	Travel objective	2001	USA	GIS map matching	Combining GPS location data with a spatially accurate GIS land use database to obtain the travel purpose: 78% of the survey cases have a good recognition effect
		2003	Germany	Multi-assignment matching clustering	A GPS data set is helpful in travel behavior analysis, and the complexity of the algorithm can be improved in the future to improve data quality
		2007	USA: California	Classification tree and discriminant analysis model	The results show that the recognition rate of the travel purpose is 62% and the overall average recognition accuracy is 73%
		2009	Netherlands	Machine learning method	Travel mode identification: the accuracy rate of car travel is the highest (75%), followed by bicycle travel (72%) and walking (68%) Travel purpose identification: the home destination has the highest accuracy rate, 74%; the store destination has a relatively low accuracy rate, 35%; other destinations have a low accuracy rate

(continued)

Table 2.3 (continued)

Data type	Research content	Year	Place	Technical mode	Main achievements
		2010	USA: New York	Iterative hierarchical matching clustering MNL model	The independent variables in the model are divided into time, historical dependence and land use variables. The matching rates of home-based travel and non-home-based travel are 67% and 78%, respectively
		2013	USA: Cincinnati region	Map matching	The accuracy of using only GPS data to identify the travel destination is approximately 59%; the accuracy can be improved to 67% by adding additional information
		2013	USA	Machine learning algorithm	The final classifier and all three input modules have the highest accuracy, 73.37%; the accuracy of a classifier based on the basic data and the travel history data of the respondents, travel history data and the travel destination is 60.57% and 72.30%, respectively. The location of the travel endpoint has the strongest influence on travel purpose correction, and the accuracy of the classifier is higher than 20%
		2014	Switzerland: Zurich	Random forest algorithm	Under the condition of ensemble operation, the recognition accuracy of the random forest is 80% and 85%
	Field application of resident travel surveys	2003	USA: California	Statistical analysis	Aiming at addressing travel underreporting, which seriously affects the results of trip generation models in resident trip surveys, this paper analyzes the impact of travel underreporting on vehicle mileage (VMT) and travel time estimation. The analysis shows that CATI obviously overestimates travel times compared with a GPS survey
		2007	Australia: Sydney	Statistical analysis	Based on an analysis of GPS survey data, it is found that approximately 7% of trips are not reported by the respondents
		2010	USA: Denver	Statistical analysis	Based on survey data of resident travel obtained by the GPS method, according to the model link classification and region type, this paper describes the characteristics of family travel in an area and supplements the regional performance data for a congestion management scheme

(continued)

Table 2.3 (continued)

Data type	Research content	Year	Place	Technical mode	Main achievements
		2011	Israel: Jerusalem	Statistical analysis	Verifies that the PR survey method based on GPS data can meet the requirements of travel data without interruption, missing data, overlaps, or a high-level space and time allocation rate
		2012	USA: Cleveland, Ohio	Statistical analysis	The travel purpose and mode identification results obtained by GPS are verified by filling in a travel log online in order to explore the feasibility of GPS traffic survey technology completely replacing the traditional questionnaire survey
		2014	USA	Statistical analysis	Based on GPS data, this paper summarizes the most valuable methods in data quality evaluation, data processing and privacy protection, and it provides methods and technical guidance for the identification of the travel mode, destination and other travel attributes
		2015	USA: Oregon	Statistical analysis	It is verified that a GPS survey is very effective in addressing the nonresponse phenomenon of young people and large families under traditional survey methods. Based on GPS data, higher travel rates, higher average vehicle occupancy and longer travel times can be identified

Table 2.4 Analysis of individual activity characteristics based on cell phone Wi-Fi data

Data type	Research content	Year	Place	Technical mode	Main achievements
Analysis of individual activity characteristics based on mobile Wi-Fi data	Analysis of individual activity characteristics	2005	USA: MIT campus	Statistical analysis	The contour map and thermal map of Wi-Fi intervention frequencies are drawn to show the main activity areas on the campus, which can guide future community planning. 55% of the people have laptops and use wireless network cards in a specific building. An average of 25% of people use Wi Fi during specific periods of the day and can estimate the number of people in the building
		2005	USA: Dartmouth campus	Markov prediction of second-order backstepping	This method has an average accuracy of 72% for users with a long tracking length, and the influential factors are discussed, such as the aging of surrounding information
		2012	Portugal	Map matching based on Wi-Fi data	By overlaying the dynamic mapping data of Global System for Mobile Communications (GSM) and Wi-Fi access points, a comprehensive analysis of the mobile network and distance map can yield more information than a single analysis of the mobile network map
		2012	Portugal: University of Minho	Map matching based on Wi-Fi data	Accurate analysis of the spatial and temporal distribution of Wi-Fi users as well as the flow between different campuses

Table 2.5 Analysis of individual activity patterns based on social network data

Data type	Research content	Year	Place	Technical mode	Main achievements
Analysis of individual activity rules based on social network data	Travel characteristics of residents	2011	England: Leeds	Agent-based model	This paper explores the effect of using emerging social network data from Twitter to evaluate a dynamic city model and proposes that semantic data mining and model application based on dynamic social data will be the focus of future research
		2012	Italy	Geographic visualization technology	This paper reveals the hot spots of urban activities as well as the travel activities and change characteristics of urban and suburban areas in different seasons, and it determines the relationship between different regions
		2015	China: Taipei	Machine-learning method	Semantic data cannot reasonably explain travel behavior, so they need to be combined with user photos for visual analysis. The SURF (accelerated feature extraction) algorithm more easily recognizes the object contours, has greater computational efficiency, and is more suitable for large-scale image analysis
	OD estimation	2014	USA: Central Chicago	Fusion of clustering, regression and gravity models	It is noted that social network data have the advantages of being available in real time and dynamic, and the data quality is high
	Spatial characteristics of occupation and residence	2015	China: Shanghai	Density-based spatial clustering algorithm	This paper analyzes the commuting travel and activity occupation residence mode contained in the location data, and the best recognition effect of "residence" is 88.3%. The work can easily overlap with other activities, so recognition is difficult
		2015	China: Chongqing	Statistical analysis	Using a large sample and a large amount of public open data can solve practical problems in the analysis of occupation and residence

In 2006, Ratti and others at MIT cooperated with Italian telecom operators to carry out real-time projects in Milan and Rome [1, 2]. The projects collected telephone and text message data through mobile phone base stations, analyzed the distribution characteristics of urban residents' activity intensity, and combined mobile phone data with GIS geographic information to explore the internal relationship between the communication base station service intensity and the regional resident density. The results showed that there is a positive correlation between individual activity intensity and base station service intensity.

(1)   Milan Citizen Activity Mode

Figure 2.1 shows the density of mobile phone calls in different areas of Milan from 9 am to 1 pm. The red and blue areas represent high-density and low-density areas of mobile phone calls, respectively. The figure reflects the commuting situation of people in Milan through transfers between high-density and low-density areas, which can effectively help decision-makers formulate corresponding traffic policies [2].

(2)   Analysis of the State of Rome during Hot Events

The project shows the mobile information of users recorded during hot events in Rome, obtains the 3D visual connection of communication strength at different time

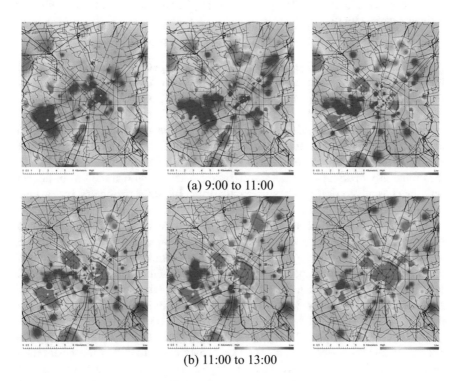

(a) 9:00 to 11:00

(b) 11:00 to 13:00

Fig. 2.1   Map of mobile phone call density in different areas of Milan

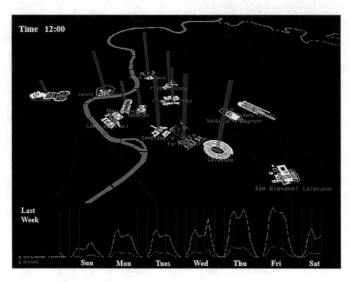

Fig. 2.2 Analysis of the attraction of tourist destinations

points in a specific area of Rome, and clarifies the movement and distribution of residents in the city [2].

(3)   Analysis of the Attraction of Passenger Flow at Rome Landmarks

Figure 2.2 shows the mobile phone usage density of people in different places of interest in Rome [2]. The curve at the lowest end of the figure is the result of comparative analysis of a week of data between the hottest area and the coldest area [3].

In 2008, Marta C. Gonzalez et al. analyzed the spatial–temporal positioning data of 100,000 anonymous mobile phones for six months, used the existing research results of biological behavior theory combined with mathematical statistics to fit the probability distribution of the travel distance, and found that the law of human activities has a high degree of spatial regularity in the same period and that people have strong regularity in travel distance and travel destination [3]. Travel patterns can be characterized by a probability distribution in space, so the similarity of individual travel has an important impact on transportation planning and other fields. Figure 2.3 shows the spatial probability distribution pattern of human travel trajectories fitted by mobile phone spatiotemporal location data, and different parameters correspond to different probability distributions. The relevant research results were published in Nature in 2008. This study showed that the clustering of individual mobile location data can provide empirical data for the basic theory of travel activity behavior [4].

In 2010, AHAS of Tartu University in Estonia noted that the number of mobile phones owned by residents, the extent of network coverage and the positioning accuracy of mobile phones have reached a high standard, and mobile phones can be used in research on combining geographic information. The analysis results based on GIS technology can be used to understand the law of urban activities in order

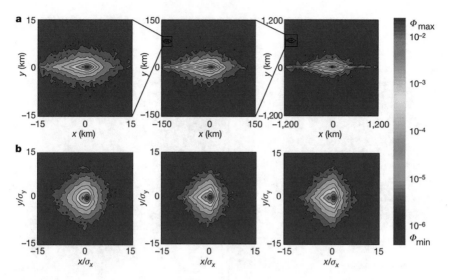

Fig. 2.3 Spatial probability distribution pattern of travel activity trajectories based on the spatiotemporal data of group mobile phone locations

to better plan and design urban space, understand the relationship between travel behavior and related factors, and better configure public service facilities [4]. This paper analyzes the characteristics of the urban population's spatiotemporal movement behavior between 2007 and 2008 by using active and passive location data and defines this method of investigating urban activities as a social positioning method. Taking Tallinn, the capital of Estonia, as the practice city, this paper studies the commuting behavior of suburban residents through the measured mobile phone positioning data, including the spatial and temporal distribution of personal location points; the main starting, ending, and stopping points of the respondents (the staying time is more than 2 h); the survey distance from the center of the city; and the daily cumulative number of trips of the respondents [5]. In addition, Estonia applies mobile positioning technology in the analysis of the tourism behavior of tourists from neighboring countries in the Tallinn area [6].

In 2014, Calabrese the United States explored the feasibility of using mobile location data to analyze the daily travel activities of large-scale urban residents through long-term location tracking of millions of pieces of mobile phone data in Singapore [7]. The study also compared the pedestrian travel mode and motor vehicle travel mode. Compared with traditional survey methods, mobile location survey technology has obvious advantages in dynamic data acquisition and data accuracy improvement. Figure 2.4 shows the comparison results of the pedestrian travel mode and motor vehicle travel mode based on travel distance analysis.

In 2014, Lai Jianhui of the Beijing University of Technology analyzed the spatiotemporal distribution of traffic demand based on mobile positioning information combined with the weekly working day data of mobile phone users in Beijing

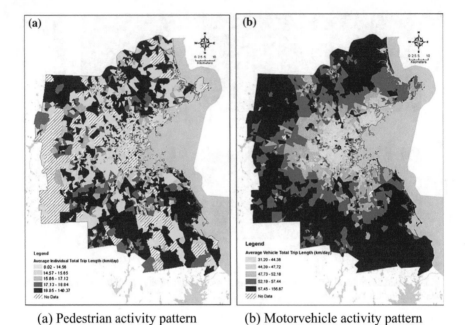

(a) Pedestrian activity pattern   (b) Motorvehicle activity pattern

Fig. 2.4 Comparison results of the pedestrian activity mode and motor vehicle activity mode based on travel distance analysis

[8]. He also analyzed the job residence distribution characteristics of rail travel users and the spatiotemporal characteristics of population flow.

(1)   Distribution Characteristics of Occupation and Residence

This paper uses mobile data to identify the residence and employment place of Beijing residents and uses a weighted average regional center to solve the problem of ping-pong switching. The spatial distribution relationship between the resident population and the employed population is shown in Fig. 2.5.

(2)   Temporal and Spatial Characteristics of Population Flow

Taking Beijing CBD staff as the research object, this paper uses mobile information to obtain the change value of the regional population and calculates the global and local Moran's index to obtain the aggregation characteristics of the regional population flow.

In 2014, Ren Ying of the Wuxi Municipal Planning Bureau carried out a travel survey based on mobile signaling data to analyze the long-term historical travel trajectory and spatial activity range of residents [9]. This research is helpful to identify the residence and employment place of residents and obtain the distribution law of employment and residence in the core area of Wuxi. The result is shown as Fig. 2.6.

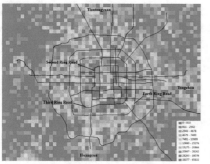 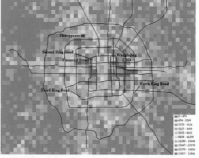

(a) Distribution of residences in Beijing   (b) Distribution of employment places in Beijing

Fig. 2.5 Distribution of residential and employment places in Beijing

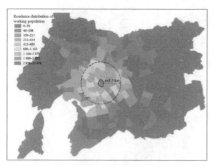 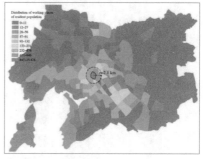

(a) Distribution of the working population in the core area
(b) Residence distribution of the working population in the core area

Fig. 2.6 Distribution of the working place and residence of the population in the core area

In 2005, Yang Dongyuan of Tongji University used the signaling data of Shanghai mobile phones to identify the residence by the time threshold method, information entropy method and time relative value discrimination method according to the different activities of residents in the daytime and at night [10]. The results of these three discrimination methods were compared, and it was found that the information entropy method had better results and a higher recognition rate.

In 2016, based on the fifth Comprehensive Transportation Survey in Shanghai, Chen Huan, from the Shanghai Urban and Rural Construction and Transportation Development Research Institute of China, analyzed the distribution of the population, occupation locations, and residences, the characteristics of commuter and tidal traffic, and the characteristics of rail transit passenger transfer [11].

(1)   Population Distribution and Occupation Residence Ratio Distribution

The call intensity of mobile phones in a specific area in a certain period of time was analyzed, to explore the law of residents' activities, and obtains the distribution of users during the day and at night. The project also analyzes the characteristics of the job and residence distributions of urban residents, which can be used as an important auxiliary means of performing demographic and employment distribution surveys during the survey period, as shown in Fig. 2.7.

2) Commuting and Tidal Traffic Characteristics

Based on daytime and nighttime data, considering the distance between the base station location and the actual road, the travel distance of the working population is obtained. The use of mobile communication data can reflect the tidal traffic characteristics of all traffic modes, thus improving the traditional survey method, which can only obtain the information of one traffic mode, as shown in Fig. 2.8.

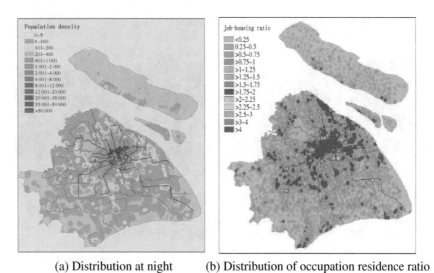

(a) Distribution at night   (b) Distribution of occupation residence ratio

Fig. 2.7 Distribution of mobile phone users at night and the ratio of workplaces to residences in Shanghai

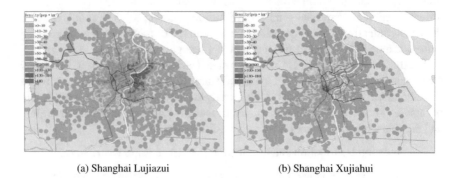

(a) Shanghai Lujiazui   (b) Shanghai Xujiahui

Fig. 2.8 Residence distribution of the working population in specific areas

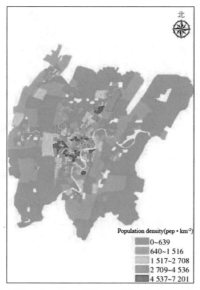
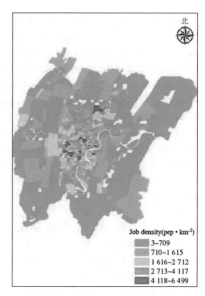

(a) The density of the residential population   (b) The density of employment

Fig. 2.9 Employment density of the resident population in the main urban area of Chongqing

3) Transfer Characteristics of Rail Transit Passengers

Based on the principle of signaling updates when a passenger's mobile phone crosses different location areas during transfer, the switching sequence of mobile phones is matched with the subway mobile phone number, and the information of passengers transferring between rail lines is obtained through a model algorithm.

In 2016, Zhou Tao and Tang Xiaoyong of Chongqing Transportation Planning and Research Institute combined Chongqing mobile signaling data to identify the residence and workplace of residents, selected evaluation indicators, and evaluated the work/residence balance of each group in Chongqing from the perspectives of independence, work/residence quantity balance and work/residence space mismatch, providing a feasible technical means for analyzing residents' work residence level [12]. In the same year, the travel characteristics of residents in different regions and different periods were studied and analyzed. After obtaining mobile phone samples of the residential population and employment in the main urban area, the sample was expanded [13]. Figure 2.9 shows the distribution of residential and employment density in the main urban area of Chongqing based on mobile phone signaling data.

2.3.2 Regional and Cross-Section Passenger Flow Analysis

1. Travel OD

In 2009, Lin Qun of the Shenzhen Urban Traffic Planning and Research Center pointed out that land use characteristics can be divided into urban traffic districts by using mobile phone signaling data [14]. The corresponding OD data can be obtained by mapping to the traffic districts according to the location changes of mobile phone users within the base station. This can provide support for urban transportation, construction, planning and management.

In 2013, Liu of Belgium analyzed the internal relationship between the individual travel purpose and traffic behavior decision-making by using the mobile phone tracking data of 80 local volunteers for more than one year while maintaining the communication mode. In the complete mobile communication mode, the call location and time are recorded so that the research can reproduce the user's communication time to analyze the call location trajectory. The results showed that the prediction accuracy of individual travel activity based on mobile signaling data can reach approximately 70% [15].

In 2013, Ran Bin of Southeast University, used the fourth comprehensive traffic survey in Tianjin and mobile phone list data provided by the operator to extract the travel information of residents and analyzed the distribution of the population residence, population post distribution, commuter travel characteristics, and OD distribution of daily travel between regions. Figure 2.10 shows the results of the OD survey in the large Tianjin area [16].

In 2014, Iqbal and others in Bangladesh constructed a resident trip OD matrix by using the call detail record (CDR) data of 2.87 million users in Dhaka city of Bangladesh for more than one month and a small amount of traffic survey data in 13 key areas of Dhaka city [17]. In this paper, the "base station to base station" OD matrix was established by analyzing the data of specific time windows in different time periods, and then it was transformed into a "node–node" OD matrix by combining the matrix with the traffic network nodes. Finally, the resulting matrix was scaled up

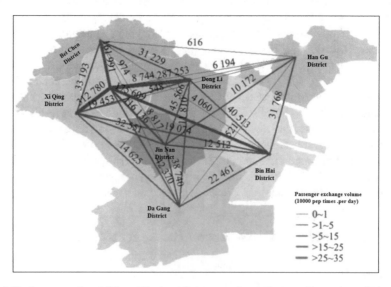

Fig. 2.10 Survey results of OD mobile data of passenger flow in large traffic sections of Tianjin

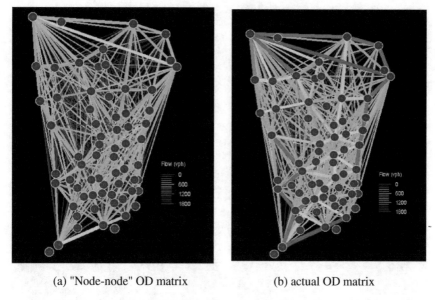

(a) "Node-node" OD matrix   (b) actual OD matrix

Fig. 2.11 Comparison results of the "node–node" OD matrix and actual OD matrix

to the actual OD matrix. Based on the optimization method, combined with the micro-scale traffic simulation platform used to determine the best matching proportion factor with the traffic survey data, the application effect was good. Figure 2.11 shows the comparison result between the "node–node" OD matrix and the actual OD matrix.

In 2014, Ren Ying of the Wuxi City Planning Bureau carried out a mobile travel survey based on mobile signaling data. According to the residence time of mobile phone users in different places in Wuxi City, the users' stay points, stop points and travel behavior were analyzed, the residents' travel OD within each area of the city was identified, and the OD was obtained by distributing the OD passenger flow to the contact network between units [9]. Based on the analysis of the passenger flow corridor, the characteristics of passenger flow in Wuxi were obtained, as shown in Fig. 2.12.

In 2016, Wu Xiangguo of the Chongqing Institute of Transportation Planning and Research, incorporating the maintenance and upgrading project of the Chongqing comprehensive transportation model, analyzed the OD matrix of commuting trips obtained by analyzing mobile phone signaling data in the main urban area of Chongqing in order to analyze the trip distribution characteristics of various groups. Figure 2.13 shows the commuter travel expectation line in the main urban area of Chongqing. It can be seen from the figure that commuter travel has no obvious radial centripetal characteristics, and most of the trips are distributed between adjacent groups. This reflects the development mode of multicenter clusters in Chongqing and reflects the accuracy of OD data obtained from mobile signaling data to a certain extent [13].

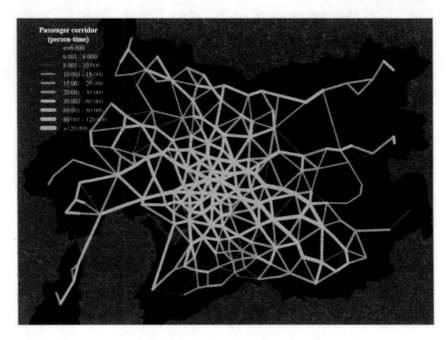

Fig. 2.12 Passenger flow corridor between urban management units on weekdays

Fig. 2.13 Commuter travel expectation line in Chongqing

2. Section Flow

In 2005, on the basis of a clustering analysis of mobile phone traffic, Li Xiaopeng and others at Tsinghua University combined the clustering results by using GIS and analyzed and divided urban activities and land use by using the time distribution characteristics of mobile phone traffic. In addition, we can obtain the pedestrian volume in time and space according to the traffic volume and quantitatively transform the traffic volume into the active population, which has certain significance in traffic planning, traffic evaluation and ITS information collection [18].

In 2015, Song Lu of Southeast University used the signaling data of mobile phone users in Shanghai and the inner ring line as the research section to obtain the OD volume of traffic in and out of the section at the morning and evening peaks [19]. The research results showed that the traffic flow in and out of the inner ring line at the morning and evening peaks on weekdays presents an obvious tidal phenomenon, and it was noted that using the signaling data of mobile phones can be used as a reference for traffic planning and management.

2.4 Individual Travel Behavior Analysis Based on Mobile Phone Sensor Data

2.4.1 Trip Chain Information Extraction

1. Travel Mode

In 2004, Nielsen et al. in Denmark, invited seven interviewees to collect outdoor GPS data for a week and fill out an electronic questionnaire [20]. The study used handheld GPS positioning equipment combined with electronic questionnaires to explore the characteristics of individual travel times, spatial movement and pedestrian travel. Studies have shown that positioning marks will be lost with the occlusion of buildings, and the GPS positioning trajectories of different traffic modes have different features. Figure 2.14a, b compare the real route and GPS trajectory of bicycles and cars. The two methods show clear differences in the density of the fixed GPS points, which is directly proportional to the velocity of the travel mode itself.

In 2006, Tsui et al. in Canada, carried out GPS field tests and simulation experiments in Toronto. They analyzed the GPS travel data of 9 volunteers over 58 days, including 103 activities, 109 trips and 237 modes. Through the experiments, the GPS data characteristics of different transportation modes were deeply analyzed. Furthermore, a fuzzy logic algorithm was used to test the recognition effect of the transportation mode using only GPS data and GPS + GIS data. The research conclusions were that compared to using GPS data alone, the use of GPS + GIS data can further improve the recognition accuracy of multiple transportation modes, such as walking, bicycles, and buses [21]. The recognition effect is shown in Table 2.6. Although the processing time of GPS + GIS data is relatively long, this research

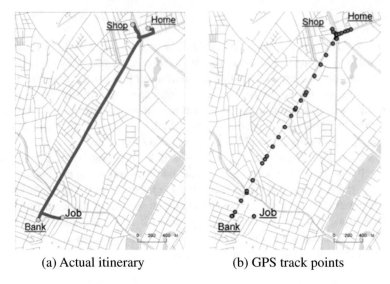

(a) Actual itinerary   (b) GPS track points

Fig. 2.14 Comparison of the actual itinerary of a bicycle trip in a street canyon with stops on the route and the GPS track points registered on the trip

Table 2.6 Travel mode detection rates: versions 1 and 2

	Version 1	Version 2	
		Detection rate	
	Average detection rate (%)	Type I (%)	Type II (%)
Walk	97	98	98
Cycle	86	72	86
Bus	76	80	80
Auto	97	97	99
Streetcar	–	88	88
Overall	91%	91	94

highlights the advantages of interactive analysis systems. Additionally, it provides a novel analysis method for GPS-based multimode travel surveys.

In 2008, Bohte et al. in the Netherlands, conducted a week-long GPS positioning data collection and analysis experiment on 1200 residents in Amersfoort, Veenendaal and Zeewold [22]. Through the integration of land use property data, GPS logs and GIS geographic information data, the process and methods of identifying individual travel modes and travel purposes were explored. On this basis, the researchers further developed an interactive website confirmation link to improve the accuracy of data recognition. Figure 2.15 shows the web page information confirmation interface of the project development.

In 2010, Manzoni et al. in America, developed data acquisition software based on a mobile phone GPS positioning chip. The research analyzed the characteristics

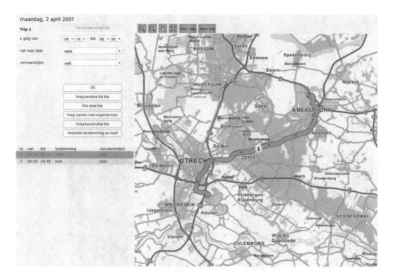

Fig. 2.15 A screenshot of the user interface of the web application

of travel data for different transportation modes and identified five transportation modes—walking, bicycle, car, bus and subway—through a rule-based algorithm. In this experiment, the average detection and recognition rate of various traffic modes reached 82.14%, among which the recognition accuracy of walking was the highest, reaching 95%; second was the recognition accuracy of cars, reaching 91.25% [23]. On this basis, the study further evaluated the carbon emissions of different transportation modes. Figure 2.16 shows the data collection and display interface of the project.

In 2012, the Ohio Department of Transportation conducted GPS travel data collection (2012) on 4,545 households in Cleveland, Ohio. The method was based on a comprehensive traffic survey, and 1,300 households were asked to fill in online travel logs. This travel log information was mainly used to verify travel purposes, travel modes and other recognition results [24]. This comprehensive traffic survey in Cleveland explored the feasibility of GPS traffic survey technology completely replacing traditional questionnaire surveys.

In 2013, Timmermans et al. in the Netherlands, collected 80,670 sets of individual travel data in London. Then, related travel logs were recorded, and the impact of GPS data and acceleration data on traffic mode recognition accuracy was studied. The experiment used a Bayesian network model to analyze the accuracy of traffic mode recognition in three cases: GPS data only, acceleration data only, and a combination of GPS data and acceleration data. The results of the comparative research found that adding acceleration data to GPS data can increase the recognition accuracy of buses and trains by 10–20%, which is a more obvious effect than for other modes of transportation [25]. However, the experiment considered only a single mode of transportation, which is not fully applicable to multimode travel.

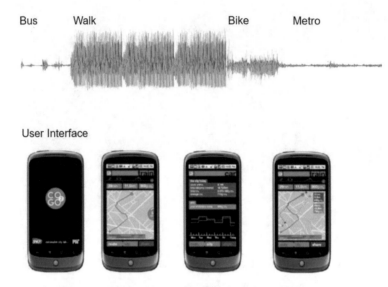

Fig. 2.16 Individual travel data collection and travel trajectory interface

In 2015, Lari et al. in Iran, conducted data collection in Tehran for more than two weeks and collected 24,518 GPS trajectories to compare and analyze the application effects of neural networks, decision trees, random forests and other algorithms. The research noted that when GPS positioning data are used to identify individual travel modes, the four factors of the algorithm mechanism, sample size, identification parameters and model assumptions have the most significant impact on the accuracy of traffic mode recognition [26]. In this experiment, the random forest model had the highest recognition accuracy for the three travel modes of bus, car and walking, reaching 87.93%, 97.68% and 90.33%, respectively. However, for certain modes of transportation with similar velocity, the detection accuracy of cars during peak hours is not high.

In 2015, Geurs et al. conducted a three-year smartphone test in the Netherlands and developed a smartphone app called *Move Smart*. The application software uses a sensor module (GPS, Wi-Fi, gravity sensor, and community ID information) to detect the user's departure and arrival time, travel start and end points, transportation mode and travel purpose. The research results showed that *Move Smart* can correctly detect most travel lengths and temporal and spatial distributions. As an alternative to traditional log survey methods, it can reduce the interviewee's recording burden and improve the accuracy of measurement [27]. However, 20–25% of trips in this app are not detected, and the battery power consumption of the mobile phone is also relatively large. Therefore, the travel mode detection rate and battery consumption efficiency still need to be improved.

2.   Travel OD

In 2007, Du et al. in America, collected 12 sets of GPS travel data (with a 10—day cycle) in Lexington and tried to combine the changes in trajectory direction, stay time,

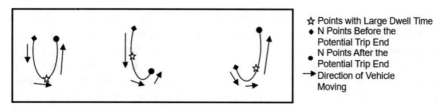

Fig. 2.17 Three scenarios of heading change around a trip end

GPS point location and road network matching relations for travel endpoint identification analysis [28]. As a result, this method effectively improved the accuracy of travel endpoint identification based on GPS data. Studies have pointed out that the setting of related parameters may significantly affect the accuracy of travel information estimation, and this method has the possibility of being optimized. Figure 2.17 is a schematic diagram of the possible direction changes of three travel endpoints.

In 2013, Bierlaire et al. in Switzerland, integrated the GPS data tracks of 25 smartphones and proposed a probabilistic map matching algorithm based on GPS sparse positioning data [29]. Then, the integration of mobile phone Bluetooth, wireless base stations, internal mobile phone gyroscopes and other types of magnetometers was proposed to extract residents' travel chains. Figure 2.18 compares the travel trajectory obtained by this method with the real trajectory.

In 2013, Zhang Jianqin et al. in Beijing, collected mobile phone location data from 8 volunteers for 20 days. Additionally, he proposed a method for identifying an individual travel status based on threshold rules, which addressed mobile phone signaling data drift issues in a large range (1–2 km) [30]. This research laid a reliable foundation for further mining and analyzing the travel chain information of permanent residents, including the number of trips, travel distance and travel time.

In 2015, Lei et al. in Japan, merged the travel data of 30 volunteers for five consecutive weeks and proposed a two-step method to identify moving points and staying points [31]. Compared with the previous density-based spatial clustering

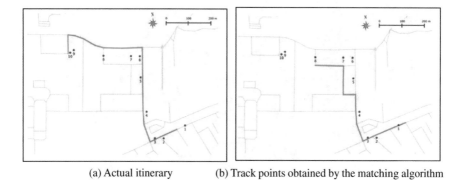

(a) Actual itinerary   (b) Track points obtained by the matching algorithm

Fig. 2.18 Comparison of the actual itinerary and matching algorithm

method, the accuracy was further increased. This research used an improved spatial clustering algorithm to identify moving points and staying points with time series restrictions and direction restrictions. Additionally, it used support vector machines to distinguish between active staying points and inactive staying points and effectively identified travel endpoints. This method still maintains good recognition accuracy even when GPS data are diversified and GPS data features are insufficient or have missing items. Figure 2.19 shows the distribution of trajectory points on the three eigenvector coordinate systems set by the support vector machine.

3.   Travel Purpose

In 2001, Wolf et al. in America, proposed the idea of combining travel destination locations and GIS data to identify the primary travel purpose [32]. The study screened out the GPS travel data and travel logs of 19 respondents for analysis. The study combined GPS location data with spatial GIS land use data to obtain travel purposes and tested the use of GPS data loggers to collect personal vehicle travel data. The research showed that the accuracy of the data obtained through the combination of GIS and traditional travel log GPS data processing methods is higher than that of traditional resident travel surveys. The collection of new travel data elements (route and velocity) makes the verification, calibration and updating of the traffic allocation

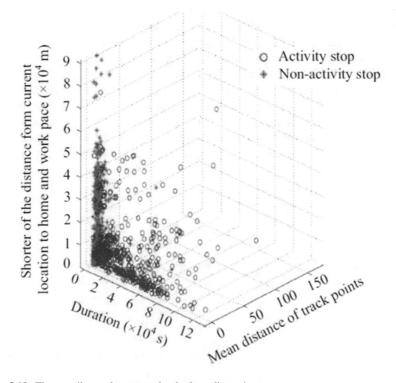

Fig. 2.19 Three attributes shown together in three dimensions

model more accurate and can improve the accuracy and completeness of the existing demand model. In addition, considering some issues, such as data errors caused by insufficient GPS positioning accuracy, missing data and oversimplification of travel detection logic, the experiment also demonstrated the effect on identifying travel purposes. The results showed that 78% of the survey cases had a good identification effect.

In 2005, Schonfelder et al. in Germany, obtained the GPS data of vehicles through a traffic safety project and used a multilayered matching procedure to identify the purpose of travel [33]. The most important task was to identify the deterministic relationship between the surveyed person's travel endpoint and demographic data to comprehensively identify the travel purpose. The study divided the investigators' travel activities according to the law of time and frequency and then used GIS to cluster the travel stops and initially determined the cluster center as a possible travel destination. On this basis, the analysis was carried out in combination with the land use of the cluster center and the predefined activity purpose. The predefined travel purpose was mainly defined by auxiliary information, such as occupation, working days/weekends, travel start time, and activity duration. The results showed that certain types of travel purposes (such as private business, work, shopping, and leisure) could be effectively identified.

In 2007, McGowen et al. in America, combined GPS data with GIS data to predict travel modes [34]. The data were a travel survey data set of California residents from 2000 to 2001. The data set includes the two-day travel logs of 17,000 families. These data mainly include GPS data collected on individuals or vehicles, land use data (geographical location type) and demographic data. This study used the classification tree method and the discriminant analysis model to identify the most likely travel purpose. Based on the combination of geocoding data, a recognition model was proposed. The model uses a large data set to determine the type of each location. The study selected five categories and a total of 26 samples of different activity types for identification. The results showed that the travel purpose recognition rate for the selected 26 types of activities was 62%, and the overall average recognition accuracy was 73%.

In 2009, Bohte et al. in the Netherlands, proposed a GPS-based travel data survey method that combined GPS data, GIS data and interactive web programs to improve the traditional GPS-based travel survey method, and a survey was conducted in the Netherlands [35]. In large-scale practice, a total of 1104 respondents participated in the one-week survey. The results showed that the recognition accuracy rate of small- and medium-sized car travel was the highest, at 75%, followed by bicycles (72%) and walking (68%). Regarding identifying the travel purpose, since the home location was known, the accuracy of travel identification with the home as the destination was the highest, 74%. The proportion of correctly identified locations was 35%, which was mainly due to the complexity of the list of store locations. For trips to socialize and visit friends, although the addresses of friends, colleagues or customers were unknown, 11% of these trips could be identified through data training and learning from previous trips.

In 2010, Chen et al. in America, used GPS and GIS technology in residents' travel surveys and studied the feasibility of using this technology [36]. The researchers asked respondents to wear clothes with GPS collectors for daily activities and used GIS-based data collection methods to collect their travel routes and travel destinations. First, a spatial multimodal transportation network database for New York City was established. An iterative method was then used to combine multiple standard algorithms to reduce the impact of the urban canyon effect. Then, the researchers used hierarchical matching clustering to cluster the travel purpose according to the OD. Finally, two multinomial logit (MNL) models were developed, one for home—based travel and the other for non—home—based travel. Through the above methods, the independent variables in the studied model were divided into time, historical dependence and land use. The matching rates of home-based travel and non—home—based travel were 67% and 78%, respectively.

In 2013, Stopher et al. in Australia, used GPS survey data in the Greater Cincinnati area of the United States to amend travel purpose identification through additional information such as frequent haunt areas, activity occurrence time, activity duration and the accuracy of travel purpose detection [37]. The study carried out a one-year data collection in the Greater Cincinnati area. The survey data were divided into five categories: homecoming, work, education, shopping and other. After analyzing the results of the 4133 GPS travel data points in the database, it was found that the accuracy of using only GPS data to identify the purpose of travel was approximately 59%, and the accuracy could be improved to 67% by adding additional information for correction. When considering additional information, such as frequent haunt locations and travel time, the recognition accuracy of the travel purpose can be improved.

In 2013, Lu et al. in America, proposed a method of using machine learning to identify travel purposes from GPS data [38]. This method combines GPS personal travel route survey records, GIS data and transportation network information to identify the purpose of personal travel. The study mainly analyzed the impact of different types of input variables, different land—use coding methods and different travel purpose classifications on travel purpose recognition. Then, it explored the feasibility of using machine learning methods to automatically identify travel purposes. The study identified five major categories of travel purposes: home—based work travel, home—based shopping travel, home-based social entertainment travel, other home—based travel and non—home—based travel. These were subdivided into ten categories: family travel, work travel, shopping travel, nursery travel, catering travel, driver travel, service travel, school travel, social travel and entertainment travel.

The data collected from the University of Minnesota based on an onboard GPS included a sample learning decision tree with 3188 trips. The final classifier and all three input modules achieved a maximum accuracy of 73.37%, where the accuracy of the surveyed basic data and travel history data was 60.57% and the accuracy of the travel purpose classifier based on the travel history data and travel destination was 72.30%.

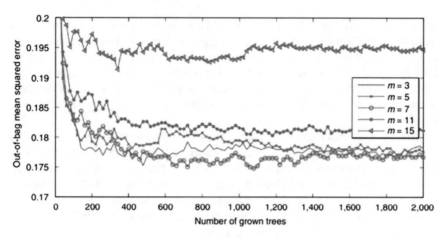

Fig. 2.20 Out-of-bag error for numbers of randomly selected features (m)

In 2014, Montini et al. in Switzerland, conducted a one—week travel data survey of 156 respondents in Zurich. Based on the respondents' GPS trajectory and acceleration data over multiple days, they studied the use of random forest algorithms in travel purpose identification [39]. The research was conducted in four stages— extracting activities and their locations, clustering locations, calculating features, and learning and applying classifiers—and it was found that under the collective operating conditions, the accuracy rate of the random forest for the purpose of travel was 85%. The research showed that the training set and its input change characteristics are very important and that different classifier settings may obtain different results. Figure 2.20 shows the different effects of the selection of the out-of-set error parameter m on the characteristics of random selection.

2.4.2 Resident Travel Survey Application

In 2003, Wolf et al. in America, used the sample data collected in a resident travel GPS survey in three areas of California to compare the same family's computer-assisted telephone calls [40]. Their aim was to discover the travel underreporting that affected the travel generation model results in the resident travel survey. The impact analysis of underreporting on vehicle mileage and travel time was carried out based on these two kinds of data. The analysis showed that computer-assisted telephone surveys overestimate the travel time and the number of trips more than GPS surveys.

In 2007, Stopher et al. in Australia, studied the accuracy assessment of residents' travel surveys based on GPS data [41]. Previous related studies showed that the traditional computer-assisted telephone survey method used in America causes 20–25% travel information loss and that GPS data can help improve the accuracy. The author selected 50 families and obtained the travel GPS data and travel log information of these families. Then, he used GPS data combined with maps or forms to complete the residents' travel inquiry surveys. It was found that the respondents missed approximately 7% of the travel information, which is a large difference compared to the traditional computer-assisted telephone survey in the United States. However, there are still some problems: missing travel distances and exaggerating the number of travels.

In 2010, Bachman et al. in America, used GPS-based resident travel survey data (437 households) to establish a velocity matrix of free flow and congestion [42]. The resident travel survey data or GPS survey data obtained by the research team could describe family travel characteristics in the area. These characteristics can supplement congestion management plans with regional empirical, livability and liquidity data. Additionally, they can provide advice to planners and decision-makers and can guide congestion mitigation strategies.

In 2011, Oliveira et al. in America, conducted a GPS—based reminder recall survey in Jerusalem, Israel, to support the development of activity—based models [43]. The development based on the activity model requires all-day travel information that is uninterrupted and complete. The current GPS—based reminder recall survey method is considered to be the most promising method of meeting these requirements. The traditional resident travel survey conducted in Jerusalem, Israel, in 2010 confirmed the feasibility of the GPS—based reminder recall method. The results showed that the GPS—based reminder recall method can provide complete and correct personal daily travel records and can mitigate the problem of not truthfully reporting travel information in traditional resident travel surveys.

In 2012, Stopher et al. in Australia, conducted two large-scale GPS-based resident travel surveys in Ohio and the local Department of Transportation in the state and explored the feasibility of GPS surveys completely replacing traditional questionnaire surveys [44]. For the first time, Stopher and Collins conducted a GPS-based family travel survey in Cincinnati, Ohio. The survey involved more than 2,000 families, of which 601 families completed a one-day reminder recall study. The results of the study showed that it is feasible to use GPS—only resident travel surveys. The second survey was conducted by GeoStats in Cleveland, Ohio, and collected information from nearly 4,000 households using a GPS to provide travel information. Of these, approximately 1,300 households participated in the reminder recall interview using computer—assisted reporting or telephone—assisted surveys. The research showed that reminder recall samples are helpful for calibrating and verifying algorithmic models that require an input of keystroke attributes.

In 2014, Wolf et al. in America, described the advancement of GPS—based resident travel survey technology in various regions of the United States and noted the advantages of smartphone GPS—based resident travel surveys, such as the low survey cost and convenient data collection [45, 46]. The report systematically summarized the research on traffic travel behavior based on GPS data since the 1990s, covering data acquisition, data formats and related algorithm evaluation methods, including GPS hardware and software, data formatting, data import and processing,

GPS equipment and related products. The researchers also summarized the most valuable methods in data quality evaluation, data processing (multiple data fusion, travel OD, travel mode, travel purpose, job and residence analysis, etc.) and privacy protection. This work provides solid guidance for travel mode and purpose recognition and other travel attribute recognition.

In 2015, in America, Bricka et al. as well as the local Ministry of Transportation, surveyed residents' travel in Oregon based on GPS [47]. A total of 299 households were divided into three groups and randomly selected for inclusion in the study. The first group was given a traditional resident travel survey. The second group was given a resident travel survey assisted by GPS data for recall. For the third group, only GPS data were collected and not resident travel surveys. The experiments showed that the traditional resident travel survey exhibited nonresponse phenomena among large families, low—income families and young people. Young people and nonminorities in the GPS group had a higher participation rate. The experimental data confirmed that GPS could effectively address the nonresponse phenomenon in young people. However, the nonresponse phenomenon in ethnic minorities increased due to the use of GPS, so some suitable methods of addressing this should be considered. Additionally, installing GPS facilities would be expensive, but these costs would decrease with the standardization of research and new technologies.

2.5 Activity Hotspots Analysis Based on Wi-Fi Data

In 2005, Sevtsuk et al. at MIT, undertook a project called ISPOTS. This project used dynamic mapping of more than 2800 Wi-Fi access points within MIT to describe people's activities within MIT, which enabled the discovery of key activity areas on campus and served as a guide for future community planning [48]. Figure 2.21 shows the use of wireless networks on campus; a heat map was created based on the

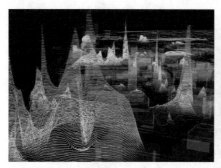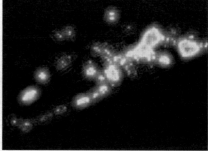

(a) Contour map of the Wi-Fi access frequency at different locations

(b) Heat map of the Wi-Fi access frequency at different locations

Fig. 2.21 Access point frequency distribution chart

number of devices accessing each location at 15—minute intervals over a 24—hour period, allowing the visualization of locations with high device access frequency. The researchers analyzed the data and concluded that 55% of the campus population has a fixed Wi-Fi usage point and 25% of the population uses Wi-Fi during specific times of the day, and the total number of people in a building can be calculated using these probabilities.

In 2005, Song et al. at Dartmouth University, concluded that location is essential in many wireless network applications [49]. The researchers used wireless networks for the first time to predict user locations and evaluated the predictions using data from nearly 6,000 Wi-Fi users on the Dartmouth campus. The study compared Markov and compression models in predicting user locations and found that the second-order backpropagation Markov model was the better prediction model. This method had an average accuracy of 72% for users with long tracking lengths.

In 2012, Conde et al. of the University of Minho, Portugal, proposed a method of studying human spatiotemporal activities based on the symbolic information of smartphone mobile networks [50]. This study analyzed the user's mobility over time by importing GSM information, Wi-Fi data, and personal mobile map information recorded in the user's mobile app into a spatial map. The study showed that importing spatial context into a mobile network map provides a better understanding of the user's daily mobility characteristics. An integrated analysis of the mobile network and distance maps yields more information than the mobile network map alone. As shown in Fig. 2.22, the location points and connection lines for GSM and Wi-Fi data are in gray, and the nodes and edges in the personal mobility map are in black.

In 2012, Meneses, at the University of Minho, Portugal, tracked each access user's mobility by using more than 550 Wi-Fi access points on campus and used this to study the state of user movement between the two campuses and downtown buildings [51]. The study analyzed the variation in the number of network users on the campus at different times of the day through time curves and compared the flow in and out

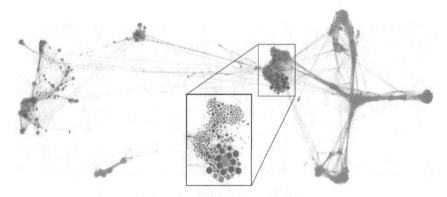

Fig. 2.22 Individual mobile roadmap overlay with Wi-Fi and GSM

between the two campuses and between the campus and the off—campus areas, taking the number of users moving between different buildings or areas to describe the closeness of the connection between buildings or areas.

2.6 Individual Travel Behavior Analysis Based on Mobile Phone Social Network Data

Traditional traffic planning is mainly based on the population scale and functional zoning for road selection and construction, and greater consideration is given to spatial layout planning at the urban planning level, ignoring the actual needs of individual residents for traffic conditions, which is less flexible. This easily causes an uneven distribution of traffic resources, traffic congestion, and other problems. With the rapid development of information technology and smartphones, users generate much social network data and check-in data in the process of using cell phones for entertainment and services, which contain much important information such as the geographic location and activity purpose. The use of data mining and visualization technology to mine massive social network data and conduct in—depth analysis and research on residents' travel and activities, thus providing powerful support for urban traffic planning and traffic management, may become a new trend in the field of transportation research in the future.

2.6.1 Resident Travel Characteristics Detection

In 2011, Birkin et al. at the University of Leeds, UK, extracted 290,215 tweets from 9223 users in Leeds over a 4—month period and constructed an agent-based model (Agent-based) to determine the basic activities of city residents and the travel patterns closely associated with them, classifying user behaviors and giving the percentage of each type of behavior [52]. Figure 2.23 shows the method of determining the basic activity behavior of users based on tweets and location point clusters [52]. This study incorporated GIS technology to visualize user analysis data and proposed the idea that the mining of semantic data and the application of models based on dynamic social data will be the focus of future research.

In 2012, Sagl et al. from the University of Heidelberg, Germany, collected user-generated mobile network traffic data (vector and overall network traffic data for targeted switching) and data from a social networking site (Flickr) and used spatial analysis to determine activity hotspots in the city as well as travel activity and the changing characteristics of urban and suburban areas in different seasons [53]. Figure 2.24 shows the density distribution of travel activity characteristics in South Tyrol in different seasons; a denser point corresponds to more frequent travel activity.

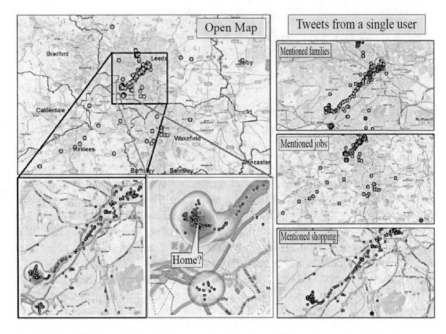

Fig. 2.23 Individual user behavior segmentation based on a Twitter keyword search

The study presented prospects for the in-depth analysis of personal travel characteristics based on social data, for example, integrated travel behavior analysis combining personal social status data and cell phone data.

In 2015, Lin et al. in Taipei, China, selected the Taipei City Zhongshan Subway Line (MRT) as the research context to capture two weeks of social data containing semantic and geographic information along the line [54]. They also used machine learning methods and computer image analysis techniques for deep mining and then obtained the activity intensity of each station. This study concluded that the photos and semantic and geographic location information uploaded by social software users would be important factors in analyzing travel characteristics. Figure 2.25 shows the analysis of station activity for the time series of the Zhongshan subway line.

2.6.2 Trip OD Estimation

In 2014, Peter et al. of Rutgers University used check-in data captured by Foursquare, a check-in-based social platform, to estimate the OD of non-commuting trips using a combination of clustering, regression, and gravity models [55]. The study compared GIS data for the central Chicago area, OD data obtained by the Chicago Department of Planning (CMAP), and check-in data collected by Foursquare. It was also noted that social network data have the advantage of being real—time, dynamic, and more

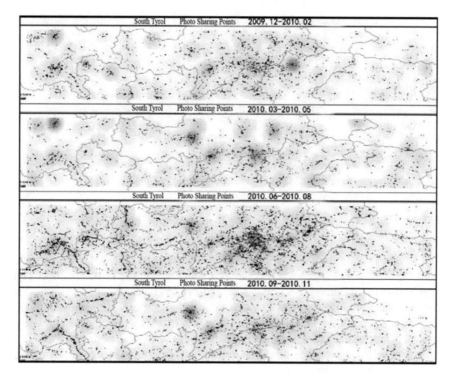

Fig. 2.24 Distribution of the travel activity density of South Tyrolean tourists in different seasons

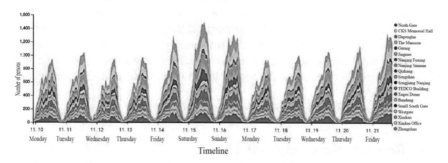

Fig. 2.25 Time series-based station activity analysis (the Taipei Zhongshan subway line is used as an example)

accurate than traditional data for transportation demand analysis, especially for OD estimation. Figure 2.26 shows a comparison between the real OD and the OD based on the check—in data.

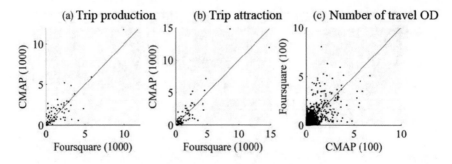

Fig. 2.26   Comparison of real data and foursquare check-in data

2.6.3   Characteristics of Job and Residence Distribution

In 2015, Mao Feng from East China Normal University collected microblog data from a total of 65,477 microblog users in Shanghai for 6 months using the Sina Open Application Programming Interface (API) [56]. A spatial clustering algorithm was used to spatially cluster all the locations posted by each user to find the resident areas of the posters, and the commuting situation and activity occupancy patterns was analyzed, embedded in the location data by fuzzy identification methods. The study showed that valuable information on occupancy and residence can be extracted from massive social network data.

In 2015, Bingrong Leng et al. of the Chongqing Municipal Planning and Design Institute used Baidu map heat map data to estimate the active population in Chongqing and further investigated the relationship between occupancy and residence [57]. The study demonstrated that the "polycentric cluster" development in Chongqing is effective, and the balance of work and residence within the clusters shortens the commuting time. This study pointed out that the use of large amounts of public open data in the era of big data can solve real problems faced in the analysis of occupancy and residence.

2.7   Research Summary and Trend

In summary, according to continuous research in many countries around the world in the past 10 years, the application of mobile positioning data to extract traffic travel parameters is still a major frontier hot spot in the international transportation field.

At present, scholars have obtained many research results on individual travel chain information extraction methods based on GPS positioning data, mobile communication positioning data and social network data; however, because of the constraints that handheld GPS survey technology equipment is expensive and the signal is easily shielded, it is still difficult to apply such equipment in urban residential travel surveys on a large scale, while cell phone communication survey technology is subject to the constraints of communication base station positioning accuracy. Usually, the positioning accuracy is tens of meters in urban areas and hundreds to thousands of meters in suburban areas. The technology is more suitable for the collection of medium—and macro—scale traffic travel characteristics and patterns, and the recognition effect is not satisfactory for the extraction of micro—scale travel chain information such as traffic modes, traffic mode interchange points and short stops.

Meanwhile, due to the large population and complex transportation environment in China, residents' daily travel is often completed through a combination of multiple transportation modes, so it is difficult for foreign data extraction algorithms to be fully applicable to Chinese cities, and for some key processing links, such as traffic travel endpoints and transportation mode interchange point identification, foreign algorithms mostly use simple travel time and distance thresholds to discriminate points, which are more empirical. Additionally, the thresholds in different studies differ significantly; for example, for travel endpoint discrimination, different studies have used 120, 180, 300 s, etc. as time thresholds, and for underground travel segment identification [23, 35, 41]; 150, 200 and 500 m have been used as distance thresholds [21, 58, 59]. These empirical thresholds are usually only valid for the specific area where the actuation is located, and when the traffic environment changes, the universal applicability of the method is not ideal. There is an increasingly urgent need in the industry for traffic parameter extraction techniques with higher data accuracy and better mining methods.

In recent years, with the rapid development of the cell phone manufacturing industry and the development of built—in cell phone GPS, accelerometers, gyroscopes and other sensor chips, and because modern smartphones are also able to obtain real—time calls, SMS, Internet access, Wi-Fi, and other rich communication network event data as well as location data from social networks, they have been integrated with GPS, accelerometers, communication network data and other positioning technology functions. With the arrival of the new generation of the 3G/4G—LTE communication era, this technology can grasp individual traffic travel characteristics more comprehensively and accurately, and the fusion application of cell phone positioning data and bus swipe cards, land use nature, and GIS geographic information data will definitely become the key to the development of future traffic travel parameter collection technology and traffic model innovation, with good prospects for professional research and industry applications.

References

1. Ratti C, Sevtsuk A, Huang S et al (2007) Mobile landscapes: Graz in real time. Location based services and telecartography. Springer, Berlin, pp 433–444
2. Ratti C, Frenchman D, Pulselli RM et al (2006) Mobile Landscapes: Using Location data from cell phones for Urban Analysis. Environ Plann B Plann Des 33(5):727–748
3. Gonzales MC, Hidalga CA, Barabasi AL (2008) Understanding individual human mobility patterns. Nature 453(7196):779–782

4. Ahas R, Aasa A, Silm S et al (2010) Daily rhythms of suburban commuters' movements in the Tallinnmetropolitan area: a case study with mobile positioning data. Transp Res Part C Emerg Technoln 18(1):45–54
5. Ahas R, Aasa A, Ular M et al (2007) Seasonal tourism spaces in Estonia: case study with mobile positioning data. Tour Manage 28(3):898–910
6. Ahas R, Aasa A, Roose A et al (2008) Evaluating passive mobile positioning data for tourism surveys: an Estonian case study. Tour Manage 29(3):469–486
7. Calabrese F, Diao M, Di Lorenzo G et al (2013) Understanding indvidual mobility patterns from urban sensing data: a mobile phone trace example. Transp Res Part C Emerg Technolo 26:301–313
8. Lai J (2014) Research on traffic information extraction and analysis method based on mobile communication positioning data. Beijing University of Technology, Beijing
9. Ying R, Mao R (2014) Mobile data and intelligent planning——taking Wuxi mobile phone data as an example. Int Urban Plann 29(6):66–71
10. Song S, Li W, Yang D (2015) Research on the method of residence identification based on mobile communication data. Integr Transp 037(012):72–76
11. Huan C, Meigen X (2016) Analysis of comprehensive traffic characteristics in Shanghai under big data environment. Urban Transp 14(1):24–29
12. Tang X, Zhou T, Liu Y (2016) Evaluation of the relationship between work and residence in the main city of Chongqing based on mobile signaling. Urban Rural Plann (Urban Geography Academic Edition) (4):10–15
13. Wu X, Yu Q, Wei C (2016) Maintenance and upgrading of Chongqing comprehensive traffic model under the background of big data. Urban Transp 14(2):51–58
14. Lin Q, Guan Z, Yang D et al (2009) Research on urban transportation planning decision support system based on mobile data. Paper presented at International Forum on innovation and development of energy saving and new energy vehicles, Shenzhen, 11–13 Dec 2009
15. Liu F, Janssens D, Cui X et al (2016) Building a validation measure for activity-based transportation models based on mobile phone data. Expert Syst Appl 41(14):6174–6189
16. Bin R (2013) Application of mobile data in traffic survey and traffic planning. Urban Transp 11(1):72–81
17. Iqbal C, Wang P et al (2014) Development of origin–destination matrices using mobile phone call data. Transp Res Part C Emerg Technol 40:63–74
18. Li X, Yang X, Cao J (2005) Analysis of urban activities and land use characteristics based on mobile phone traffic. Paper presented at China intelligent transportation annual meeting, Shanghai
19. Lu S (2015) Research on traffic OD distribution based on mobile location data. Southeast University, Nanjing
20. Nielsen TS (2004) Henrik Harder Hovgesen. GPS in pedestrian and spatial behavior surveys. Paper presented at the 5th International Conference on Walking in the 21st Century, June 9–11, 2004
21. Tsui SYA, Shalaby AS (2006) Enhanced system for link and mode identification for personal travel surveys based on global positioning systems. Transp Res Record J Transp Res Board 1972(1):38–45
22. Bohte W, Maat K (2009) Deriving and validating trip purposes and travel modes for multi-day GPS-based travel surveys: a large-scale application in the Netherlands. Transp Res Part C Emerg Technol 17(3):285–297
23. Manzoni V, Maniloff D, Kloeckl K et al (2010) Transportation mode identification and real-time CO_2 emission estimation using smartphones. SENSEable City Lab. Massachusetts Institute of Technology, Cambridge, Massachusetts, USA, Technical Report
24. U.S. Bureau of Transportation Statistics (2012) BTS Statistical Standards Manual. U.S. Department of Transportation, Research and Innovative Technology Administration
25. Feng T, Timmermans HJP (2013) Transportation mode recognition using GPS and accelerometer data. Transp Res Part C Emerg Technol 37(3):118–130

26. Ansari Lari Z, Golroo A (2015) Automated transportation mode detection using smart phone applications via machine learning: case study mega city of Tehran. Paper presented at the 94th Transportation Research Board Annual Meeting, Washington D.C.
27. Geurs KT, Thomas T, Bijlsma M et al (2015) Automatic trip and mode detection with MoveSmarter: first results from the Dutch Mobile Mobility Panel. Transp Res Proc 11:247–262
28. Du J, Aultman-Hall L (2007) Increasing the accuracy of trip rate information from passive multi-day GPS travel datasets: automatic trip end identification issues. Transp Res Part A 41:220–232
29. Bierlaire M, Chen J, Newman J (2013) A probabilistic map matching method for smartphone GPS data. Transp Res Part C 26:78–98
30. Zhang J, Qiu P, Xu Z et al (2013) A travel itinerary recognition method based on mobile phone positioning data. J Wuhan Univ Technol 37(5):932–938
31. Gong L, Sato H, Yamamoto T (2015) Identification of activity stop locations in GPS trajectories by density-based clustering method combined with support vector machines. Transport 23(3):202–213
32. Wolf J, Guensler R, Bachman W (2001) Elimination of the travel diary: experiment to derive trip purpose from global positioning system travel data. Transp Res Record J Transp Res Board 1768:125–134
33. Schönfelder S, Samaga U (2005) Where do you want to go today? More observations on daily mobility. Paper presented at the 5th Swiss transport research conference, Monte Verità/Ascona, 9–11 Mar
34. McGowen P, McNally M (2007) Evaluating the potential to predict activity types from GPS and GIS data. Paper presented at Transportation Research Board 86th Annual Meeting, Washington D.C.
35. Bohte W, Maat K (2008) Deriving and validating trip destinations and modes for multiday GPS-based travel surveys: application in the Netherlands. Paper presented at Transportation Research Board 87th Annual Meeting, Washington D.C.
36. Chen C, Gong H, Lawson C et al (2010) Evaluating the feasibility of a passive travel survey collection in a complex urban environment: Lessons learned from the New York City case study. Transp Res Part A Policy Pract 44(44):830–840
37. Shen L, Stopher PR (2013) A process for trip purpose imputation from global positioning system data. Transp Res Part C Emerg Technol 36(36):261–267
38. Lu Y, Zhu S, Zhang L (2013) Imputing trip purpose based on GPS travel survey data and machine learning methods. Paper presented at Proceedings of the 92nd Annual Meeting of the Transportation Research Board. Washington D.C.
39. Montini L, Rieser-Schüssle N, Horn A et al, Trip purpose identification from GPS tracks. Transp Res Record J Transp Res Board 2405(-1)
40. Wolf J, Oliveira M, Thompson M (2003) Impact of underreporting on mileage and travel time estimates, results from global positioning system - enhanced household travel survey. Transp Res Rec 1854:189–198
41. Stopher P, FitzGerald C, Xu M (2007) Assessing the accuracy of the Sydney household travel survey with GPS. Transportation 34:723–741
42. Bachman W, Oliveira M, Xu J et al (2012) Household-level global positioning system travel data to measure regional traffic congestion. Transp Res Record J Transp Res Board 2308:10–16
43. Oliveira MGS, Vovsha P, Wolf J et al (2011) Global positioning system-assisted prompted recall household travel survey to support development of advanced travel model in Jerusalem Israel. Transp Res Record 2246(1):16–23
44. Wargelin L, Stopher P, Minser J et al (2012) GPS-based household interview survey for the Cincinnati, Ohio Region. Ohio. Dept. of Transportation. Office of Research and Development
45. Wolf J, Bachman W, Oliveira MS, Auld J, Mohammadian A, Vovsha P (2014) NCHRP report 775: applying GPS data to understand travel behavior, volume I: background, methods, and tests. Transportation Research Board, Washington D.C.
46. Wolf J, Bachman W, Oliveira MS, Auld J, Mohammadian A, Vovsha P, Zum J (2014) Medicine NCHRP report 775: applying GPS data to understand travel behavior, volume ii: guidelines. Paper presented at transportation research board, Washington D.C.

47. Bricka S, Zmud J, Wolf J, Freedman J (2015) Household travel surveys with GPS—an experiment. Transp Res Record J Transp Res Board
48. Sevtsuk A, Huang S, Gutierrez D et al (2005) How wireless technology is changing life on the MIT Campus. http://senseable.mit.edu/ispots. Accessed on 14 Sept 2005
49. Song L, Kotz D, Jain R et al (2005) Evaluating next-cell predictors with extensive WiFi mobility data. IEEE Trans Mob Comput 5(12):1633–1649
50. Pérez-Penichet C, Conde Â, Moreira A (2012) Human mobility analysis by collaborative radio landscape observation. In: Proceedings of the AGILE' 2012 international conference on geographic information science, Avignon, pp 24–27
51. Filipe Meneses, Adriano Moreira (2012) Large scale movement analysis from WiFi based location data. In: International conference on indoor positioning and indoor navigation, pp 13–15
52. Birkin M, Malleson N (2011) Microscopic simulations of complex metropolitan dynamics. Paper presented at the 5th-6th Complex City workshop, Amsterdam
53. Sagl G, Resch B, Hawelka B (2012) From social sensor data to collective human behaviour patterns—analysing and visualising spatio-temporal dynamics in urban environments. Geovizualisation Soc Learn 2012:54–63
54. Lin C, Chen AY (2015) Trip characteristics study through social media data. Paper presented at international workshop on computing in civil engineering, Austin, 21–23 June 2015
55. Jin PJ, Cebelak M, Yang F (2014) Location-based social networking data: an exploration into the use of a doubly constrained gravity model for origin-destination estimation. Paper presented at transportation research board, Washington D.C.
56. Feng M (2015) Research on residents' commuting behavior and spatial characteristics of urban occupation and residence based on multi-source trajectory data mining. East China Normal University, Shanghai
57. Leng B, Yu Y, Huang D et al (2015) Analysis of the relationship between work and residence in Chongqing's main urban area from the perspective of big data Planner
58. Chung EH, Shalaby A (2005) A trip reconstruction tool for GPS-based personal travel surveys. Transp Plan Technol 28(5):381–401
59. Chen C, Gong H, Lawson CT et al (2010) Evaluating the feasibility of a passive travel survey collection in a complex urban environment: lessons learned from the New York City case study. Transp Res Part A Policy Pract 44(10):830–840

Chapter 3
Methodology for Mobile Phone Location Data Mining

With the widespread popularity of smartphones, the phenomenon of "one phone for everyone" has become a reality. Mobile phone sensors can record a large number of accurate location data of the traveler, including time, GPS latitude and longitude, velocity, acceleration, and base station communication data. Massive and accurate mobile phone location data can provide accurate sources of resident travel information for transportation planning and decision—making. However, the individual travel information (including the starting point and ending point, mode of transportation and travel purpose) cannot be obtained directly from the mobile phone location data, and the original mobile phone location data can only be processed by the corresponding data mining algorithm to extract the individual travel chain information. Therefore, the reasonable selection of a data mining algorithm is the key issue that is introduced in this chapter.

The main research content of individual trip chain information extraction based on mobile phone location data is the trip starting and ending point (OD), transportation mode and trip purpose recognition; this research is in the category of pattern recognition [1, 2]. The common data mining algorithms in pattern recognition include statistical analysis, cluster analysis, machine learning and image recognition. In this chapter, common recognition data mining algorithms such as cluster analysis, machine learning and image recognition are comprehensively used to process mobile phone location data to extract accurate individual travel chain information and analyze the applicability of these algorithms.

3.1 Technology Structure for Individual Travel Chain Information Extraction

Individual mobile phone location data cannot directly show individual trip chain information, so a corresponding data mining algorithm is needed to interpret them [1]. Individual travel chain information mainly includes the departure point (OD

F. Yang and Z. Yao, *Travel Behavior Characteristics Analysis Technology Based on Mobile Phone Location Data*, https://doi.org/10.1007/978-981-16-8008-3_3

information), the mode of transportation and the purpose of travel. Among them, the identification of the departure point is the basis of the identification of the transportation mode and travel purpose [3]. Extracting individual travel chain information through a data mining algorithm is the key problem in performing fine mobile traffic surveys. At present, many scholars are trying to use efficient and high—extraction—accuracy algorithms to identify travel endpoints, transportation modes and travel purposes. ① For OD identification, a spatial clustering algorithm based on density is mainly used to identify the travel endpoints with high aggregation density. ② The identification of the traffic mode is mainly based on the change characteristics of travel data such as velocity and acceleration, and the wavelet analysis and machine learning algorithm combined with geographic information technology is used to identify the traffic mode. ③ Travel purpose identification is mainly based on individual travel habits combined with the corresponding land use nature.

The refined extraction of individual travel chain information from individual mobile phone location data can effectively compensate for the problems of the unclear recall of respondents and low quality of the survey data in traditional residential travel surveys.

The individual mobile phone location travel chain data are usually a complete day or more than one day of travel location data, which include the information of multiple trips by individuals, and each trip also contains information on various modes of transportation. Therefore, as shown in Fig. 3.1, the extraction ideas for individual trip chain information in this chapter are as follows: first, the densely based spatial clustering algorithm is used to identify the trip endpoint information, and the individual mobile phone location trip chain data are divided into multiple single trip data according to the trip endpoint. Second, according to the mobile phone location data of each trip, the wavelet analysis algorithm is used to identify the transfer points of traffic modes in a trip, and the data of different traffic modes are divided into the transfer points of a trip. Finally, a machine learning algorithm (neural network, support vector machine, Bayesian network, or random forest) combined with GIS data (site matching algorithm) is used to identify different traffic modes, such as walking, bicycle, car and bus, according to the data characteristics of different traffic modes [4]. The results show that the travel chain information extraction method proposed in this chapter has good operability and high identification accuracy.

3.2 Trip End Recognition Based on Spatial Clustering Algorithm

In a broad sense, activities of residents going out are called travel. However, from the perspective of traffic influence and traffic planning, it is considered that a trip from A to B has the following three characteristics: ① A purposeful activity is completed. ② A street or highway with a name is used. ③ The one—way walking time is more than 5 min, or the one—way cycling distance is more than 500 m (this value may

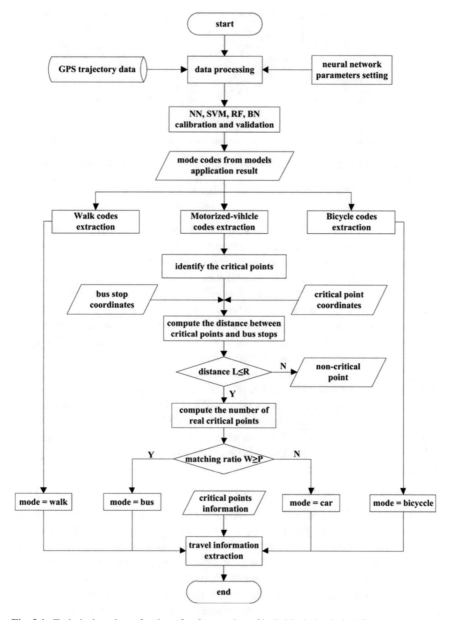

Fig. 3.1 Technical roadmap for the refined extraction of individual trip chain information

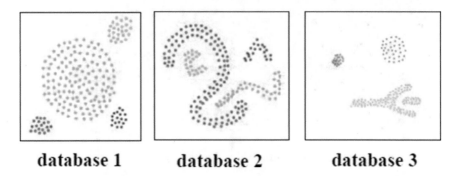

database 1   **database 2**   **database 3**

Fig. 3.2 DBSCAN spatial point clustering effect diagram

vary according to the size of the urban space) [5]. Because residents will carry out activities with a specific purpose at the beginning and end of each trip, that is, they will stay for a long time, the corresponding mobile phone location trajectory will form a dense point cluster, which is quite different from the relatively sparse scattered points along the journey. The spatial clustering algorithm based on density can be used to make the trajectory points obviously clustered, and the travel endpoints with a high density of point space can be identified in the form of point clusters. On the other hand, the track points are scattered, and a travel route with a low point space density will be identified as a set of noise points. Therefore, the spatial density of the track points can better determine whether the individual mobile phone location travel chain data belong to the trip route or the trip endpoint. At present, density—based spatial clustering algorithms mainly include DBSCAN, OPTICS, and DenClue [6]. (The clustering effect is shown in Fig. 3.2.) The core idea of these clustering analysis algorithms is to cluster all data points according to the spatial density of the track points. If a point cluster is formed, it is identified as a trip endpoint; if not, the points are identified as a trip route.

1.   Basic Principle of the Spatial Clustering Algorithm

The spatial clustering algorithm based on density is the process of defining the density of the core point, taking the core point as the starting point, constantly expanding into the surrounding areas according to the density requirements and finally forming the point cluster. The formation of point clusters means that travelers stay for a long time in a small range and the track points are relatively dense; they can generally be regarded as travel endpoints. Density—based spatial clustering of applications with noise (DBSCAN) is a classical method of identifying point clusters effectively [7]; however, since the definition of the core point by this algorithm does not consider the time attribute of the track points, as shown in Fig. 3.3 (1), the recognition effect is not ideal for the three cases of missing GPS signals resulting in missing data, repeated paths and multiple travel endpoints in the same place. Therefore, for a time series of track points in a point cluster, the algorithm can be better used to identify the travel endpoint information. Spatial—temporal DBSCAN (ST—DBSCAN) comprehensively considers the spatial—temporal distance between track points to

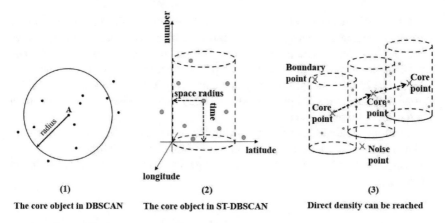

(1)   (2)   (3)

The core object in DBSCAN   **The core object in ST-DBSCAN**   **Direct density can be reached**

Fig. 3.3 Core objects, boundary objects, and noise

improve the definition of core points. Clusters are defined as maximum sets of points that are densely connected, and the algorithm can divide regions with sufficiently high density into clusters and find clusters of any shape in a spatial database with noise. The detailed definitions are as follows:

E neighborhood: The region of a given object with a spatial distance radius of Eps and a time distance of ΔT is the E neighborhood of the object.

Emphasis: If the sample points in the field of a given object E are greater than or equal to $MinPts$, the object is described as having emphasis, as shown in Fig. 3.3 (2).

Direct density can be reached: For a sample collection D, if the sample points q and p are in domain E, and p has emphasis, then from object q, the density of p can be reached directly, as shown in Fig. 3.3 (3).

Density—reachable: For sample set D, given a series of sample points P_1, P_2, ... P_n, where $P = P_1$ and $Q = P_n$, if object P_i is directly densifiable from P_{i-1}, then object Q is densifiable from object P.

Density connection: If there is a point O in the sample set D, and if object O is densifiable from object P to object Q, then densities P and Q are related.

The density—reachable set is the transitive closure of the direct density—reachable set, and the relationship is asymmetric. The density connection is symmetric. The purpose of DBSCAN is to find the largest collection of density—connected objects.

Using ST—DBSCAN to process individual travel GPS trajectory data, three parameters, Eps, the ΔT neighborhood and the corresponding sample points $MinPts$ in this field, are defined. The three—parameter concordant rule determines the space—time density of the core point and the corresponding expansion conditions of the point clusters.

2.   Travel Endpoint Recognition and Travel Track Cutting Based on a Spatial Clustering Algorithm

Using a spatial clustering analysis algorithm based on density to process mobile phone GPS positioning data and extract travel endpoint information requires three steps. The first step is data preprocessing. The second step is cluster analysis and travel endpoint identification. The third step is to organize the travel schedule.

First, due to GPS signal obscurity or mobile phone performance problems, the original mobile phone GPS positioning data need to be removed to provide correct data for travel endpoint identification.

Second, the spatial clustering algorithm based on density is used to carry out spatial clustering, cluster the point clusters, and identify travel endpoints.

Finally, the travel chain is cut again and again according to the travel endpoint information.

(1) Data Preprocessing

The mobile phone GPS signal will be weak when the city canyon effect or buildings block it. Under weak conditions, the GPS track points will often have a large position deviation. Four visible satellites are a prerequisite for accurate positioning, so the positioning record is deleted if the number of visible satellites is less than four. The velocity threshold value of the track point and the threshold value of the distance between adjacent points (such as the velocity limit value of the road section) are set, the location records such that the instantaneous velocity and distance between adjacent points are greater than the set threshold values are deleted, and the error caused by the mobile phone operation problem is reduced. Finally, because signal loss data are more common when entering a room or passing through a viaduct, underground tunnel, etc., rules can be set to supplement the data under the condition of GPS signal loss: ① Calculate the distance between the time track points before and after signal loss. If the distance is greater than the set value, it can be assumed that there is a long period of displacement when signal loss occurs, which is mainly due to travel under tunnels, subways, viaducts, etc. In this period of data, there are few travel endpoints (and if there are travel endpoints, it is difficult to identify them by using only mobile phone sensor data), and supplementary points are not considered. ② If the distance is less than the set value, the missing signal is usually for a short time or due to entering a room. The points can be filled in evenly according to the positions of the points before and after the missing signal and the data length.

(2) Trip Endpoint Identification

To identify the travel endpoints by using the spatial clustering algorithm based on density, the corresponding parameters Eps, ΔT and $MinPts$ should be calibrated first. A cumulative probability distribution analysis is first performed for the trip endpoint stay duration of the sampled trip chain data of various travel destination types, and the trip endpoint stay duration with a cumulative probability value reaching a certain percentage is $MinPts$. Second, the approximate distance of the $MinPts$ value of each track point is calculated, a line chart is drawn to find the inflection point in the line chart (the difference in curvature before and after the inflection point is large), and the distance of the inflection point is used as the parameter calibration value of Eps. Considering the time series when defining the core point—that is, the condition of core point expansion is to satisfy both the space distance and the time distance requirements—we have $\Delta T = MinPts$.

After parameter calibration for the clustering algorithm is completed, the algorithm is used to carry out clustering analysis on other trip chain data. The identified point cluster is the trip endpoint, the points contained in the point cluster are the trip endpoint stay points, and the start—stop span length is the trip endpoint stay length.

(3)   Travel Path Cutting

If the interval time between two travel endpoints is less than a specific value, it indicates that the individual did not make a trip, and a combination of travel endpoints is needed. By sorting out the travel endpoint information, the complete mobile phone GPS travel chain trajectory is divided to form multiple single travel trajectories under the principle that the end time of each travel endpoint is the end time of the last trip, namely, the beginning time of the next trip.

3.3 Mode Transfer Point Recognition Based on Wavelet Analysis Algorithm

The wavelet transform modulus maximum algorithm has an outstanding ability to identify the main features of nonlinear structural data and detecting the singularity of signals. It also provides good local information in the frequency domain and time domain of signals, which provides an important means of describing the singularity of signals. It is an efficient signal data processing method [8].

Cities in China, especially large cities such as Beijing, Shanghai, and Chengdu, have larger urban areas, great travel distances, and limited single transportation method coverage; therefore, people usually use a variety of traffic mode combinations, such as bus transfer to the subway and bike transfer to buses, so transportation transfer point recognition presents a difficult problem for data acquisition.

1.   The Principle of the Wavelet Transform Modulus Maximum Algorithm

A singularity in signal data has been defined in detail in the field of mathematics [9]. If the signal $f(x)$ has a mutation at a certain point or the derivative is discontinuous, it is said that the signal has a singularity at this point. The Lipschitz index (*Lip* index) is generally used to describe the singularity degree of the signal $f(t)$, and a nonnegative integer n is set, $n \leq a \leq n+1$, if there is a constant $A > 0$ and $P_n(t)$. For $t \in (t_0 - \delta, t_0 + \delta)$, the polynomial $P_n(t)$ of degree n yields:

$$|f(t) - p_n(t - t_0)| \leq A|t - t_0|^{\alpha} \tag{3.1}$$

$f(t)$ is *Lip* α at t_0. If the *Lip* exponent α of $F(t)$ at t_0 is less than 1, then t_0 is a singularity of $f(t)$, and then the wavelet transform coefficient at this point can yield the modulus maximum. Therefore, the modulus maximum points of the wavelet transform system can also reflect the singularity of the detected signal, which is especially suitable for processing nonstationary signals. The relationship between the modulus maximum points and signal singularities is as follows:

Under a certain scale a_0, if there exists a point (a_0, b_0) such that

$$\frac{\partial W_f(a_0, b_0)}{\partial b} = 0 \qquad (3.2)$$

then the point (a_0, b_0) is a local extreme point. If any point b in a certain field of b_0 is given, then we have:

$$\left| W_f(a_0, b) \right| \leq \left| W_f(a_0, b_0) \right| \qquad (3.3)$$

Then, (a_0, b_0) is a modulus maximum of the wavelet transform. The connection of all the modulus maximum points in the scale space (a, b) is called the modulus maximum line.

In theory, the location identification of singularities depends on the intersection of the maxima lines of positive and negative moduli at low scales. The maximum value line of positive and negative modulus can be extended to the point where the scale is close to zero to ensure that the two curves intersect in order to accurately find the position of the signal singularity. In practical applications, the key is to select the appropriate wavelet analysis function so that the positive and negative modulus maximum lines intersect at a point so that the location of the singularity can be accurately identified.

2.   Identification Method and Process of Determining Transportation Transfer Points

When using the wavelet transform modulus maximum algorithm to identify the transfer points of different traffic modes, two steps are needed: first, singularity identification is performed; second, the identification results are denoised. First, because of the obvious data characteristics of different modes of transportation, singularities occur when two different modes of transportation meet. Second, for some data signals containing noise, the identification results based on the modulus maximum algorithm may contain a large number of nontransfer singularities, so it is necessary to reduce the identification error of the transfer points by means of denoising steps.

(1)   Singularity Recognition

For the identification of transportation transfer points, when resident travel transportation changes, the velocity of the GPS instrument is recorded; for example, singularity signals will exhibit sharp fluctuations, the moment has a modulus maximum of wavelet transform coefficients corresponding to continuous wavelet transform scale changes, and transportation point mutations in the time—scale plane will produce at least one modulus maximum line perpendicular to the timeline. In addition, according to the principle of wavelet transform, when the analysis scale is small, the mother wavelet will oscillate rapidly and decay rapidly, and the time positioning ability will be strongest. However, with the increase in the analysis scale, the wavelet will expand, the oscillation frequency of the mother wavelet will slow, and the time positioning ability will worsen [10]. Therefore, small scale wavelet transform modulus maximum points for resident transfer times are the most accurate transportation positioning data, and the transportation transfer points in this study can be obtained through the smallest—scale wavelet transform modulus maximum to accurately extract the corresponding time points; namely, the modulus maximum line and time axis intersection for the traffic mode change to these points in time.

However, as mentioned above, a singularity corresponds to at least one modulus maximum line in the time—scale plane, which means that it is possible to have a singularity corresponding to multiple modulus maximum lines, and the number of modulus maximum lines depends on the order of the vanishing moment of the wavelet function. Therefore, the appropriate selection of the wavelet function will have an important effect on singularity recognition. For the wavelet function $\psi(t)$, if for all $k < n$, we have

$$\int_{-\infty}^{+\infty} t^k \psi(t) dt = 0 \tag{3.4}$$

Then this means the wavelet function has nth—order disappearing *torque (t)*. The number of modulus maximum lines increases linearly with increasing vanishing moment order. Too many modulus maximum lines will lead to the confusion of the singularities, so it is not suitable to use the wavelet function with an excessively high vanishing moment to detect singularities. However, the order of the vanishing moment of the wavelet function is also related to the size of the detected singularity *Lip* exponent. When the maximum value of the detected singularity *Lip* exponent is n, the adopted wavelet function should have a vanishing moment of at least order n. Therefore, the principle of selecting the wavelet function should be to select the wavelet function with the least vanishing moment on the premise that the maximum *Lip* index value can be detected.

For residents traveling with a combination of multiple modes of transportation, the transfer of modes of transportation can usually be considered to be instantaneous at a certain location. Therefore, when choosing the wavelet function, the vanishing moment of the function should be as small as possible, preferably first order. At the same time, the selection of the wavelet function should take into account the accuracy of the time identification of singularities.

(2) The Recognition Results Denoising

On a small scale, using the wavelet modulus maximum algorithm to identify a change in the mode of transportation is the most accurate method, but when there is strong background noise, such as a vehicle stopping at an intersection or congestion, the algorithm easily mistakenly identifies a car—to—walking transition. Phenomena such as GPS trajectory points and secondary feature signals or noise in signal waves are very similar; therefore, it is necessary to apply multiscale wavelet transform modulus maxima under changing properties for noise reduction.

We depict the signal singularity index of the Lip_α and wavelet modulus maximum $|W_\alpha f(x)|$ that exists between the following relationships: assuming that $\alpha \leq 1$ or less, the function $f(x)$ on $[a, b]$ has a consistent Lip index, and there is a constant $k > 0$ such that $x \in [a, b]$, the wavelet transform satisfies:

$$|W_a f(x)| \leq ka^\alpha \tag{3.5}$$

Take the logarithm of both sides of Eq. (3.5) to obtain:

$$\log|W_a f(x)| \leq \log k + \alpha \log a \tag{3.6}$$

where α is the wavelet transform scale. According to Eq. 3.6, if the Lip index $\alpha > 0$, then the wavelet transform modulus maximum of the function will increase with increasing scale; if $\alpha < 0$, then the wavelet transform modulus maximum of the function will decrease with increasing scale. Because the signal index is generally greater than zero and the index of the noise of the $Lip\alpha$ tends to be less than zero, the signal and noise under different scales of the wavelet transform present the opposite characteristics. According to this feature, singular point denoising can be realized; namely, the removal amplitude decreases with the increase of the scale of the points, and the preservation amplitude increases with the increasing scale of the points.

Therefore, when the wavelet transform modulus maximum algorithm is used to identify the transfer points of traffic modes, the wavelet transform scale should be set to a larger value to accurately obtain the singularity and the modulus maximum line of the change points of traffic modes. At the same time, the interference caused by secondary features or noise, such as fluctuations due to parking and velocity changes at intersections, should be eliminated. On this basis, wavelet transform scales should be chosen as smaller values due to the small scale modulus maxima of singular point times for the most accurate location; therefore, the time points corresponding to the master modulus maximum line (or the modulus maximum line with timeline nodes) are the resident transportation transfer points in time for performing data noise reduction.

3.4 Travel Mode Recognition Based on Machine Learning Algorithm

Machine learning (ML) is an interdisciplinary field involving probability theory, statistics, approximation theory, convex analysis, algorithm complexity theory and other disciplines. Its focus is on how computers can simulate or implement human learning behaviors to acquire new knowledge or skills and how to reorganize existing knowledge structures to continuously improve their performance. It is the core of artificial intelligence and a basic way to make a computer intelligent, and it is applied in all fields of artificial intelligence; it mainly uses induction and synthesis rather than deduction [11].

Machine learning algorithms are an important method of pattern recognition. A popular topic for scholars around the world has been to use them to identify traffic modes [12]. The daily travel modes of residents mainly include walking, bicycles, buses, cars and subways, and each mode of transportation has its corresponding travel characteristics, such as that walking velocity is commonly 5–10 km/h and bicycles are stable at 10–20 km/h; car and bus travel features are similar, but there are generally more buses parked on streets than cars (bus parking is located in places such as bus stops, intersections, and car parking areas), and this mainly occurs at intersections. Because of underground subway travel channels, it is difficult to obtain satellite signal coverage, so GPS signals cannot be collected in these areas; if there is a travel period with only two pairs of latitude and longitude coordinates, this travel characteristic difference is very obvious. In this way, different modes of travel can present different characteristics. As long as the input attributes of a reasonable machine learning algorithm are determined, the machine learning algorithm can be used to automatically interpret and identify different modes of transportation and combination information.

3.4.1 Neural Network Algorithm

The backpropagation (BP) neural network is one of the main applications of pattern recognition; it can use a meaningful GPS data input vector associated with a specific mode of transportation and can identify travel by looking for different modes of transportation and output data. Regarding the relationship between transportation and network training, training can be used in the model for automatic identification. The BP neural network consists of three modules: an input layer, hidden layer and output layer. It can continuously learn and train the potential relationship between the input and output data pairs in a hidden layer between the pair. In subsequent feedback, the trained network can then predict the output (mode of transportation) based on the newly entered GPS measurements of the phone [13, 14].

1. Basic Principle and Structure of the Neural Network Algorithm

The BP network is a multilayer forward—propagation network that is trained by an error propagation algorithm. It is currently one of the most widely used neural network models. The BP network can learn and store a large number of input—output pattern mapping relations without revealing the mathematical equations describing the mapping relations beforehand. Its learning rule is to use the fastest descent method to continuously adjust the weight and threshold of the network through backpropagation to minimize the sum of the squared error of the network.

The basic principle of a BP neural network model in processing information is as follows: the input signal X_i enters through intermediate nodes (hidden layer points), and at the output node, through nonlinear transformation, the output signal Y_k is produced. Each sample includes the training of the input vector X and the desired output t and the network output value Y and t; the deviation of the desired output is determined by adjusting the coupling strength values of the input nodes and hidden layer node W_{ij}, and the hidden layer nodes are adjusted by the coupling strength between the output node T_{jk} and the threshold. The error decreases along the gradient direction, and after repeated training, when the minimum error of the corresponding network parameters (weights and threshold) is obtained, training is stopped. The trained BP neural network can process nonlinear transformed information with the minimum output error for input information of similar samples.

The topology of the BP neural network model consists of an input layer, hidden layer and output layer. A single hidden—layer feedforward network is commonly known as a three—layer feedforward network or three—layer perceptron, namely, an input layer, middle layer (also known as the hidden layer) and output layer. The characteristics of the neural network are as follows: the neurons in each layer are only fully connected to the neurons in the adjacent layer, there is no connection between neurons in the same layer, and there is no feedback connection between the neurons in each layer, thus forming a feedforward neural network system with a hierarchical structure. A feedforward neural network with a single computing layer can only solve a linearly separable problem, and a neural network that can solve a nonlinear problem must be a multilayer neural network with a hidden layer. Since mobile phone GPS data and traffic modes are not linearly separable, it is necessary to train a multilayer neural network for recognition.

2.   Traffic Mode Identification Method Based on a Neural Network Algorithm

There are three steps when using a neural network algorithm to identify traffic modes: first, the model is established; second, the number of training layers is set; and third, traffic mode identification is carried out. First, the input layer, hidden layer and output layer of the neural network model are set to establish the relationship with the model. Second, the numbers of neural network layers and neurons are set according to the accuracy and efficiency of training recognition. Third, the trained neural network is given mobile phone GPS data input for traffic mode identification.

1)   Model Building

(1)   Input Layer

For the attribute setting of the neural network, the point velocity is the most basic distinguishing attribute under different traffic modes, and the maximum velocity per minute is also a very important distinguishing point, especially when distinguishing motor vehicles from pedestrians and bicycles. Moreover, because motor vehicles are more likely to be affected by traffic congestion than other modes of travel, the range of velocity change per minute of different modes of travel in traffic congestion also shows great differences. Finally, together with the unit displacement data collected by GPS devices, these four attributes are used as the input layer attributes of the neural network. The specific mathematical equation is:

$$X_i = \begin{bmatrix} x_1(i) \\ \cdots \\ x_N(i) \end{bmatrix} = \begin{bmatrix} v(i) \\ s(i) \\ p(i) \\ d(i) \end{bmatrix} \tag{3.7}$$

where $v(i)$ is the average velocity vector of trip I; $s(i)$ represents the maximum minute velocity vector of this trip; $p(i)$ represents the velocity variance vector of this trip; and $d(i)$ represents the combined acceleration variance vector collected by the GPS device.

$$v(i) = \begin{bmatrix} v_1(i) \\ \cdots \\ v_n(i) \end{bmatrix}, s(i) = \begin{bmatrix} s_1(i) \\ \cdots \\ s_n(i) \end{bmatrix}, p(i) = \begin{bmatrix} p_1(i) \\ \cdots \\ p_n(i) \end{bmatrix}, d(i) = \begin{bmatrix} d_1(i) \\ \cdots \\ d_n(i) \end{bmatrix} \tag{3.8}$$

The number of input neurons in the neural network model is:

$$N = n * m \tag{3.9}$$

Here, n is the number of trips in the sample, and M is the number of input attributes (the value of this model is 4).

(2)　Hidden Layer

$$H(i) = \begin{bmatrix} h_1(i) \\ \cdots \\ h_m(i) \end{bmatrix} = \begin{bmatrix} \varphi\left(\sum_{j=1}^{N} \omega_{j,1} x_j(i) + b_1\right) \\ \cdots \\ \varphi\left(\sum_{j=1}^{N} \omega_{j,m} x_j(i) + b_m\right) \end{bmatrix} \tag{3.10}$$

Here, $h_m(j)$ represents the value of the mth implied neuron; $w_{j,m}$ represents the association weight from the jth input neuron to the mth implied neuron; h^{th} represents the error between the mth implied neuron and the exact value; and φ is a conversion function. In this study, a hyperbolic tangent function with fast convergence velocity is used, as follows:

$$\varphi(y) = \frac{\exp(2y) - 1}{\exp(2y) + 1} \tag{3.11}$$

(3)   Output Layer

$$Y(i) = TT(i) = \varphi\left(\sum_{k=1}^{m} w_k h_k(i) + b\right) \qquad (3.12)$$

Here, $Y(i)$ and $TT(i)$ denote the expected mode of transportation for travel i; w_k represents the association weight of the kth implied neuron and output neuron; b stands for the output error; φ is a conversion function; and the output unit often uses a linear function.

2)   Setting the Number of Training Layers

After establishing the BP neural network model and transforming the traffic mode into data, we can start to use the neural network model to identify the traffic mode. In identifying the mode of transport, we first need to determine the neural network layer number and the number of neurons and, according to the principle of neural network training, increase the training layers. The error will reduce with neural network training, and a pattern will automatically be found that makes the relationship between the input and output attributes closer to reality; therefore, the model prediction accuracy will also rise sharply. However, when the number of training layers exceeds a certain value, the training process of the neural network will become very complex, and the training time will continue to increase. At the same time, the training results will exhibit an overfitting phenomenon, that is, overtraining, which makes the relation conditions between the input and output attributes too harsh, and the training results will not have strong applicability.

3)   Traffic Identification

Using the trained neural network model, we input the four characteristic attribute values of the mobile phone GPS data and record the output results of the neural network output layer as the traffic mode identification results.

3.4.2   Support Vector Machine Algorithm

Support vector machines (SVMs) are an important machine learning method proposed in statistical learning theory [15]. They are also the most widely used pattern classification technology with the best comprehensive effect. Traffic mode recognition based on mobile phone location data belongs to the typical category of pattern recognition, and an SVM algorithm has good applicability in identifying traffic modes. The basic idea of SVMs is to solve the classification problem between two kinds of samples and to find a hyperplane that can be divided between the two kinds of samples in high—dimensional space to achieve efficient and accurate classification of the two kinds of samples. Compared with other classification algorithms, its unique advantage is that it can distinguish linearly inseparable data sets well [16].

Due to the use of mobile phone GPS positioning data (including velocity, acceleration, velocity variance, positioning accuracy, etc.), the correspondence relationship
between transportation modes is not linear, as the same velocity value does not correspond to the same kind of transportation; common linear classification algorithms
such as logistic regression do not apply in the identification of transportation modes,
and using SVMs can better solve this problem of nonlinear classification [17–19].

1. Basic Principle and Structure of the SVM Algorithm

The basic idea of an SVM is to construct a hyperplane in a high—dimensional
space to find the maximum interval between different categories. It is determined
whether the hyperplane is linear according to the input data. Due to the differences
in individual transportation modes in many respects (velocity, acceleration, velocity
variance, positioning accuracy, etc.), nonlinear SVMs are mainly used to identify
transportation modes. Their basic principle is that the input vector is mapped to a
high—dimensional feature vector space. If an appropriate mapping function is chosen
and the dimension of the feature space is sufficiently high, most nonlinear separable
patterns in a high—dimensional feature space can be converted into linear separable
models, as shown in Fig. 3.4; therefore, they can be classified in the feature space to
construct an optimal surface in order to perform pattern classification of the structure
and inner product nuclear relation. To solve this problem, the method commonly used
by scholars is to introduce a kernel function [20].

In the case of nonlinear classification, the nonlinear problem can be transformed
into a linear problem in a high—dimensional feature space, and the optimal classification surface can be obtained in the transformation space. Moreover, the mapped
high—dimensional feature space may be finite dimensional or infinite dimensional,
and this mapping can be expressed as:

$$x \rightarrow \Phi(x) = (a_1\Phi_1(x), a_2\Phi_2(x), \cdots, a_n\Phi_n(x)), a_n \in R, \Phi_n \in R \qquad (3.13)$$

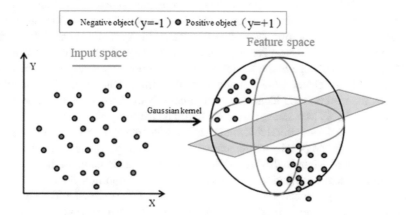

Fig. 3.4 Example graph of a SVMs with a linear separable high—dimensional feature space

In nonlinear SVMs, the separation process is mainly performed by a kernel function $K(x_i, x_j) = \Phi(x_i) \cdot \Phi(x_j)$.

2.   Traffic Identification Method Based on an SVM Algorithm

There are three steps in using the SVM algorithm to identify traffic modes. First, the feature vector and kernel function are selected. Second, the penalty coefficient c is calibrated with the kernel parameter γ. Third, traffic mode identification is performed. First, due to the obvious data feature differences between different modes of transportation, the data feature can be used as the feature vector. At the same time, the kernel function is a necessary tool to map the feature vector to the high—dimensional feature vector space, and it needs to be selected. Second, the penalty coefficient c and kernel parameter γ are calibrated to obtain the globally optimal SVM. Finally, on the basis of the trained vector machine, mobile phone GPS positioning data are input to identify the mode of traffic.

(1)   Eigenvector and Kernel Function Selection

According to the characteristic differences of the GPS data of different traffic modes, the average velocity vector, the maximum minute velocity vector, the velocity variance vector and the acceleration variance vector are selected to form the input feature vector. The Gaussian kernel, because of its flexibility, is one of the most commonly used kernel functions for mapping input parameters into a high—dimensional or infinite—dimensional plane. The mathematical expression of the Gaussian kernel function is as follows:

$$K(x_i, x_j) = \exp\left(-\|x_i - x_j\|^2 \big/ 2\sigma^2\right) \tag{3.14}$$

(2)   The penalty Coefficient c is Calibrated with the Kernel Parameter γ

The penalty coefficient c adjusts the proportion of the confidence range and experiential risk of the learning machine in the defined feature subspace to ensure that the generalization ability of the learning machine is optimal. For each feature subspace, there is at least one appropriate c value that makes the generalization ability of the SVM optimal. The Gaussian kernel parameter γ directly affects the performance of SVMs. In fact, a change in the kernel parameter γ implicitly changes the mapping function and thus changes the complexity of the sample feature subspace distribution. For the traffic mode identification problem, if the kernel parameter γ is not appropriate, the SVM cannot achieve the expected learning effect.

According to the structural risk minimization principle of the SVM, it is necessary to optimize both the penalty coefficient c and the kernel parameter γ to obtain the global optimal SVM. In statistical learning theory, many estimations of the upper and lower bounds in risk estimation are developed; a certain proportion of the GPS data in the sample (1/10) is selected by applying the method of the optimal grid system. For each of these parameters, the combination of c and γ can ensure recognition accuracy in estimating the transportation mode through comparing the different performances of transportation parameter combinations to determine the highest value (as shown

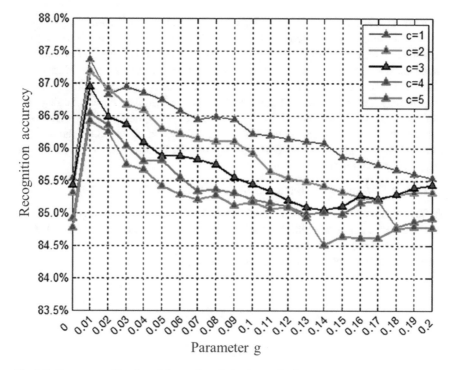

Fig. 3.5 Parameter calibration of the mesh optimization method

in Fig. 3.5). Then, the best SVM is found that is suitable for mobile phone GPS data. Therefore, the empirical risk and confidence range of the trained SVM are close to the optimal combination, and there will be neither an "overlearning" nor an "underlearning" phenomenon, so this method has good applicability.

(3) Traffic Identification

Based on the SVM trained in the previous step, the four feature vectors of mobile phone GPS data are used as input data for classification. The numbers 1, 2, 3 and 4 represent four modes of transportation, namely, walking, bicycle, bus and car, respectively. The classification and identification results are summarized and counted.

3.4.3 Bayesian Network Algorithm

A Bayesian network, also known as a belief network or directed acyclic graph model, is a probability graph model [21]. It is an uncertainty processing model that simulates the causal relationship in the human reasoning process, and its network topology structure is a directed acyclic graph (DAG) [22]. A Bayesian network based on probabilistic reasoning is proposed to solve the problems of uncertainty and imperfection. It is suitable for expressing and analyzing uncertain and probabilistic events and can make inferences from incomplete, imprecise or uncertain knowledge or information. Transportation recognition based on mobile phone GPS data can evolve into such data through the GPS feature determining the maximum probability of the transportation process. Due to the data characteristics and the uncertainty of the relationship between transportation and probabilities, using Bayesian networks for the analysis of mobile phone GPS data transportation in probabilistic inference is reasonable and has a certain popularity.

1.   Basic Principle and Structure of the Bayesian Network Algorithm

A Bayesian network is a DAG consisting of nodes representing variables and the directed edges connecting them, as shown in Fig. 3.6. The nodes represent random variables, and the directed edges between the nodes represent the mutual relationship between the nodes (the edges point from parent nodes to child nodes). The

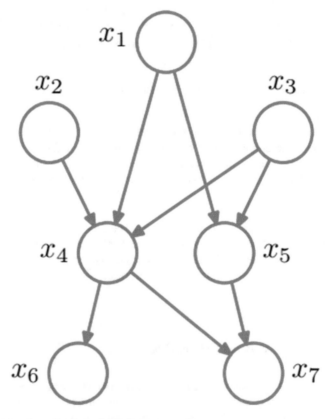

Fig. 3.6 Directed acyclic graph of a Bayesian network

strength of a relationship is expressed by a conditional probability, while information without parent nodes is expressed by a prior probability. Node variables can be abstractions of any problem. They can be used to express and analyze uncertain and probabilistic events, to make decisions that are conditionally dependent on multiple control factors, and to make inferences from incomplete, imprecise or uncertain knowledge or information.

Let $G=(I, E)$ represent a directed acyclic graph, where I represents the set of all nodes in the graph, E represents the set of directed connecting line segments, and $X=(X_i)(i \in I)$ is a random variable represented by a node i in the directed acyclic graph. Then, the joint probability assignment of node X can be expressed as:

$$P(X_1 = x_1, \cdots, X_n = x_n) = \prod_{i=1}^{n} (X_i = x_i | X_{i+1} = x_{i+1}, \cdots, X_n = x_n) \quad (3.15)$$

The Bayesian network method has the advantages of saving storage space and easily obtaining the dependence between variables or the conditional independence [23].

2. Traffic Mode Identification Method Based on the Bayesian Network Algorithm

The process of traffic mode identification based on the Bayesian network algorithm includes two main steps. The first step is Bayesian network construction. The second step is traffic mode identification. First, according to the mobile phone GPS data and the corresponding traffic mode used to build the Bayesian network, the key point is the construction of the network topology and the formation of a conditional probability table. Then, on the basis of the constructed Bayesian network, the mobile phone GPS data are input, and the probabilities of the occurrence of various modes of transportation are determined; the highest probability is the identification result.

1) Bayesian Network Construction

The construction of a Bayesian network includes two main parts: determining the topological relations among random variables and forming a DAG. The goal is to train the Bayesian network, that is, to complete the construction of the conditional probability table.

(1) Determination of the Topological Relation among Random Variables to form the Acyclic Graph

It is necessary to construct a relatively optimal topology network through iteration and improvement.

(2) Train the Bayesian Network

The conditional probability table is constructed. Since the average velocity, maximum minute velocity, velocity variance and acceleration variance can be observed directly, the training in this step is intuitive, and the method is similar to naive Bayes classification [24].

When the structure or parameters of the Bayesian network are unknown, the structure or parameters of the network must be estimated by using the observed mobile phone GPS sample data. Generally, it is more difficult to estimate the network structure than the parameters of the nodes. Since the collected mobile phone GPS data are complete but the network structure is unknown (not completed by a domain expert), the whole model space needs to be searched. Additionally, as the GPS data types are relatively singular, a general network structure satisfying parameter convergence can be found. On the basis of this network structure, the conditional probability of each node is evaluated by using the observed mobile phone GPS sample data. Generally, random reasoning (the Monte Carlo method) is adopted to determine the conditional probability: First, there will be given numerical variables that are fixed, and the other numerical variables are not arbitrarily given an initial value. Then, the iteration steps begin; after the iteration, if the first several numerical values are not yet stable, the approximate conditional probability distribution can be calculated.

2)   Traffic Identification

Based on the Bayesian network trained by using the mobile phone GPS sample data, the remaining mobile phone GPS data are input to calculate the occurrence probability of various traffic modes, and the traffic mode with the highest occurrence probability is taken as the identification result for recording and statistics.

3.4.4 Random Forest Algorithm

As the name implies, a random forest establishes a forest in a random way. There are many decision trees in the forest, and the decision trees in the random forest are not correlated. After the forest is obtained, when a new input sample comes in, each decision tree in the forest is asked to make a judgment to determine to which category the sample belongs (for the classification algorithm). Then, the most often selected category is determined, and the sample is predicted to be in that category. Traffic mode recognition based on mobile phone GPS data is a typical classification process, so it is reasonable to use a random forest algorithm to process GPS data to classify various traffic modes. At the same time, the random forest is randomized in the use of variables and data, which does not easily lead to overfitting, and it has good anti—noise ability. In addition, it can deal with data of very high dimensions (with many features) without feature selection and has strong adaptability to data sets. These advantages can ensure that the random forest has good applicability when processing mobile phone GPS data to identify traffic modes [25, 26].

1.   Basic Principles and Structure of the Random Forest Algorithm

A random forest is a classifier containing multiple decision trees. Therefore, it is necessary to be familiar with decision trees to understand random forests. Decision tree learning is one of the most widely used inductive reasoning algorithms. Its basic principle is to classify instances by arranging them from the root node to a certain leaf node, which is the classification to which the instance belongs. Each node in the tree specifies a test for an attribute of the instance, and each successive branch of the node should have a possible value of the attribute. The method of sorting instances is to start at the root node of the tree, test the attribute specified by that node, and then move down the branch corresponding to that attribute value for the given instance. This process is then repeated in the subtree with the new node as the root. Decision tree algorithms include ID3, Assistant, C4.5 and other widely used algorithms.

In 2001, Leo Breiman and Adele Cutler proposed a new decision algorithm: the random forest algorithm. The random forest constructs a decision forest in a random way, and it is composed of many decision trees. In other words, the use of variables (columns) and data (rows) is randomized to generate many decision trees, and then the results of the decision trees are summarized. There is no correlation among the decision trees in the random forest. For each decision tree, the training set it uses is derived from the total training set, which means that some samples from the total training set may appear multiple times in the training set of a tree or may never appear in the training set of a tree. When training the nodes of each tree, the features used are randomly extracted from all the features in a certain proportion without putting them back. The essence of the random forest algorithm is an improvement of the decision tree algorithm. Multiple decision trees are merged together, and the establishment of each tree depends on an independent sample. Each tree in the forest has the same distribution, and the classification error depends on the classification ability of each tree and their correlation. Feature selection uses a random method to split each node and then compares the errors generated under different circumstances. The number of features selected is determined by the inherent estimation error, classification ability, and correlation that can be detected. The classification ability of a single tree may be small, but after randomly generating a large number of decision trees, a test sample can statistically determine the most likely classification through the classification results of each tree. Therefore, the classification accuracy of the random forest algorithm is higher than that of the decision tree algorithm. The structure and decision—making process of the random forest algorithm are shown in Fig. 3.7.

2.   Traffic Mode Identification Method Based on the Random Forest Algorithm

The application of the random forest algorithm to traffic mode identification requires four steps: first, the random selection of GPS positioning data; second, the random selection of eigenvectors; third, the construction of a decision tree; and fourth, random forest voting classification. First, a GPS data set is selected randomly as the training set to generate the decision tree. Second, a certain number of eigenvectors (velocity, acceleration, etc.) are randomly selected to determine the best splitting point. Then, several decision trees are constructed according to the general construction method of decision trees. Finally, the generated decision trees are combined into a random forest for classification and identification of transportation modes.

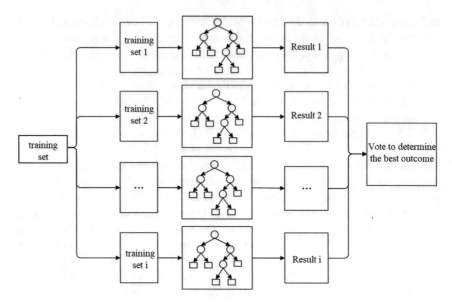

Fig. 3.7 Random forest structure

1)   Random Selection of GPS Location Data

The number of samples in the GPS data training set is N, and the N samples are obtained (by the bootstrap method) through repeated sampling with resets. These sampling results will be used as the training set to generate the decision tree.

2)   Random Selection of Eigenvectors

If there are M input variables (generally, four are selected in this chapter: the average velocity, maximum minute velocity, variance of velocity, and variance of acceleration), each node will randomly select m ($m < M$) specific variables and then use these M variables to determine the optimal splitting point. During the generation of the decision tree, the value of m remains unchanged.

3)   Building a Decision Tree

Based on the sample sets randomly generated in the first two steps, the general decision tree construction method can be used to obtain a classified (or predicted) decision tree, and each decision tree grows as far as possible without pruning. It should be noted that the random selection feature method should be used when calculating the optimal classification feature of nodes. In this model, the *Gini* index is used to determine the best attribute judgment of each node, and the split attribute selection rule is that the attribute with the minimum *Gini* index is selected as the split attribute [27].

$$gini(T) = 1 - \sum_{j=1}^{N} P_j^2 \tag{3.16}$$

where P_j refers to the classification proportion of transportation mode J at node T.

4)   Random Forest Voting Classification

By following the three steps above, a decision tree can be obtained. This process can be repeated many times (the number of repetitions is the number of decision trees), and the generated decision trees form a random forest. The random forest classifier is used to discriminate and classify new data, and the classification results are determined according to the number of votes of the decision tree. In other words, the transportation mode with the most votes (walking, bicycle, bus, or car) is selected as the result of random forest identification.

3.5   Trip Chain Information Optimization Based on GIS Map Matching

A GIS is a specific and very important type of spatial information system. It is a technical system that collects, stores, manages, calculates, analyzes, displays and describes the geographical distribution data on all or part of the Earth's surface (including the atmosphere) with the support of computer hardware and software systems [28, 29]. The GIS can match residents' travel paths with traffic road networks and analyze residents' traffic behavior patterns intuitively. At the same time, it is indispensable to check the validity of GPS data. In the method of traffic identification based on mobile phone GPS positioning data, a machine learning algorithm can better identify the walking, bicycling and driving travel modes, but the identification accuracy is not high when distinguishing buses and cars. This is because the travel characteristics of buses and cars are very similar, such as the average travel velocity, maximum travel velocity and unit time displacement, and buses and cars often queue up with each other when they travel, which further increases the difficulty of distinguishing the two modes of transportation. However, there is an obvious difference between buses and cars: buses stop at bus stops and signalized intersections, while cars usually stop only at signalized intersections. Therefore, this study incorporates GIS map matching to discriminate these two modes of traffic and proposes a stop matching algorithm based on matching key track points and bus stops.

When urban residents use motor vehicles (buses and cars) to travel, they will stop in the road network. For example, cars will wait for red lights at signalized intersections and stop fewer times at bus stations. Buses not only wait for red lights at signalized intersections but also stop at bus stops on the corresponding bus routes. GIS matching technology (a station matching algorithm) was used to match the travel stop points with bus stops and intersections. According to historical experience and experimental results, the threshold value of the station range and the threshold value of the stop percentage were set. When the distance between a stop point and a bus stop is within the range threshold value, the vehicle can be considered to stop at the bus stop. After counting the proportion of vehicles parked at each bus station and reaching the set threshold value of the stopping percentage, a vehicle can be considered a bus; otherwise, it can be considered a car. Based on this, the two travel modes can be accurately discriminated by GIS geographic matching technology [30].

1.   The Basic Principle of the Stop Matching Algorithm

The core idea of the stop matching algorithm is to match the location data of the key points of a motor vehicle travel track with the location data of a bus stop on the same track to distinguish buses and cars by the degree of matching. Therefore, the definition of the key points and the matching between the key points and the station are the key problems of this algorithm.

The key points in the travel trajectories of motor vehicles are defined as the time points at which different modes of transportation meet. A time at which one mode of transportation ends and another mode of transportation begins is called a key point in the process of traveling by combining multiple modes of transportation. At the same time, because of the phenomenon that buses and cars, in the process of traveling, will stop at bus stops and intersections, and the travel velocity at these moments is effectively zero, machine learning algorithms, due to this temporary value, cannot achieve an efficient recognition effect; buses and cars that are parked could easily be wrongly identified as the travel mode of bicycling or walking. The locations of these stops are also referred to as key points. Therefore, the key points in the process of identifying traffic modes by machine learning algorithms include two categories of key points: the key points at which the real travel mode changes and the key points that are wrongly identified due to parking and other reasons. In the following pattern recognition process, these two kinds of key points need to be distinguished to obtain the real key point information because the pattern discrimination method and the time point identification of the traffic mode change as a result. Figure 3.8 is the schematic diagram of the key points. In the figure, the transportation mode changes from 1, walking mode, to 3, bus mode. The transportation mode of the blue points changes,

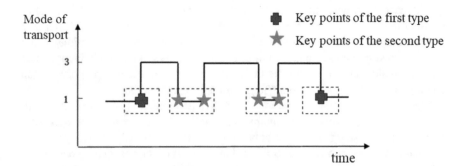

Fig. 3.8 Classification diagram of the key points

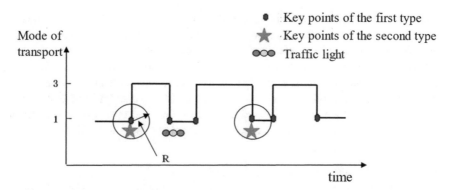

Fig. 3.9 Schematic diagram of bus station matching

and these points are the first type of key points. The red points, where events such as parking occur, are the second type of key points.

Matching between key points and bus stops: The distances L_i between the key points and each bus stop are calculated, as shown in Fig. 3.9. If the distance is less than a certain threshold R, then vehicle parking is considered to occur at a bus stop. When the proportion of stopped stations W among the bus stations along the travel path exceeds a certain value P, the mode of transportation is considered to be a bus, and otherwise, it is a car. The main idea here is to distinguish buses from cars: a bus will travel along the path of bus stops, at which at least one docking phenomenon occurs, and the dock site percentage will be greater than a certain value, but a car will not exhibit a docking phenomenon. According to the different bus and car travel characteristics, this method will be able to accurately discriminate the two modes of buses and cars.

$$L_i = \sqrt{(x_i - \hat{X}_k)^2 + (y_i - \hat{Y}_k)^2} \tag{3.17}$$

(x_i, y_i) represents the coordinates of key point i; (X_k, Y_k) represents the coordinates of an actual bus stop k.

$$W = \frac{\sum_{i}^{n} n}{N} \tag{3.18}$$

where $n_i = \begin{cases} 0, & (r_i \geq R) \\ 1, & (r_i < R) \end{cases}$, $n_i = 1$ represents that the vehicle stops at bus stop i, and N represents the total number of bus stops that the vehicle passes during the trip.

2. Optimization Method of Vehicle Travel Mode Identification Results Based on the Stop Matching Algorithm

There are three steps in optimizing the identification results of the vehicle travel mode by using the station matching algorithm: first, vehicle data preprocessing; second, site matching; and third, optimization of the vehicle identification results. First, on the basis of using a machine learning algorithm to identify traffic modes, the GPS data identified as motor vehicle traffic modes are intercepted to provide a data basis for station matching. Second, buses and cars can be distinguished effectively by incorporating the matching algorithm of key points and bus stops in the travel trajectory. Finally, the optimized results are segmented to output the model results.

1)   Motor Vehicle Data Preprocessing

The machine learning algorithm is used to identify the traffic mode, and the identification results are saved. On this basis, data processing methods such as mode data integration, abnormal point correction, iterative tensor decomposition (ITD) correction of mode transitions and unreasonable mode correction are carried out on the preliminary identification results. The specific methods are as follows:

(1)   Integration Processing: Since the data obtained through the machine learning algorithm to identify transportation modes are decimal data and the transportation modes defined are integer data (walking mode is 1, bicycle mode is 2, bus mode is 3, and car mode is 4), it is necessary to integrate the decimals first. The method here converts the data values in the range 0.5–1.5 to 1, those in the range 1.5–2.5 to 2, those in the range 2.5–3.5 to 3, and those in the range 3.5–4.5 to 4.

(2)   Abnormal Point Correction: as a result of the step treatment, most of the data are able to shift to the true target, but a few edge data points, such as 1.4 and 2.6, may truly be bicycle transportation, but may be incorrectly classified as walking or public transportation. Therefore, here, we need to modify these data, and the specific method is as follows: according to the sequence of travel times, 15 pattern recognition result points are successively taken as a processing unit, and the mode with the highest occurrence frequency among the 15 points is taken as the transportation mode of the first point. Then, the transportation mode for the whole travel time is processed, and the processing results are stored.

(3)   Modification of Mode ITD: After the above treatment, frequent abrupt changes in traffic modes can be eliminated, and each traffic mode will persist for a relatively long period of time before changes occur. However, there are still problems in the data at this time. In a transition period of the transportation mode, such as a transition from walking to car travel, the velocity gradually increases. The machine learning algorithm can easily identify a time period in which the velocity is higher than walking and lower than car travel as the bicycle mode, and this error needs to be corrected. According to the velocity variation characteristics of the ITD of the actual traffic mode and the test results of multiple data groups, this kind of unreasonable traffic mode is modified to the traffic mode at the higher end of the velocity.

(4)   Unreasonable Pattern Correction: After performing the above two steps to obtain pattern recognition results of limited wave number, data may be lacking

or there may be no real data, and adjustments are still needed to modify such data; all the kinds of transportation are summarized by the shortest travel time and distance, a certain threshold is set, and the return transportation is modified to the real situation.

2) Site Matching

For passengers at key points of traffic mode changes, that is, key points at which the vehicle stops, the latitude and longitude coordinates of the key points are imported from the original data according to the key point information obtained above. Additionally, the final coordinates of all bus stops along the travel path are imported, and the distances L_i between the key points and each bus stop are calculated. If the distance is less than a certain threshold R, then vehicle parking is considered to be at a bus stop, and when travel occurs along the path of the bus stops and the visited sites account for a proportion of all sites along the route that exceeds a certain threshold PW, the mode of transportation is considered to be a bus; otherwise, it is a car.

3) The Identification Results are Sorted and Output

According to the above steps, the bus and car modes can be accurately determined. Next, the mode results need to be output in stages. The specific methods are as follows: The key information is identified, recorded and maintained, and according to the key information, the travel process is divided into different modes of transportation. Then, the pattern recognition result is a block output of 3 or 4 for buses and cars; at the same time, according to the key information, the transportation change points are output in terms of time, location and other information.

The real key point of a transfer from walking to a bus should be at a bus stop, but due to GPS positioning error and neural network identification error, the key point will be at some distance from the bus stop. The setting of the matching distance R, and therefore the size of the threshold percentage P, will affect the final bus pattern recognition as a result; therefore, to increase the recognition accuracy of the model as much as possible and reduce the error of transportation change points, we need to set different thresholds of pattern recognition and test them to determine the appropriate matching distance R and threshold percentage P.

3.6 Summary

Through the analysis of the data characteristics of mobile phone positioning data in this chapter and the corresponding relationships among individual pieces of travel chain information, a set of complete chains of individual travel information based on mobile phone positioning data extraction methods were summarized: namely, according to the travel chain mobile phone positioning of individual complete trajectory data, a spatial clustering algorithm based on density was used to identify travel endpoint information, and according to the travel endpoint, individual travel chains were divided into more than one travel mode. On this basis, the use of the wavelet transform modulus maximum algorithm to identify each trip according to different modes of transportation and the integrated use of machine learning algorithms, neural networks, SVMs, Bayesian networks, and random forests were combined with the GIS geographic information data matching algorithm to perform accurate identification of walking, bicycles, cars, buses and other modes of transportation. The results showed that the trip chain information extraction method proposed in this chapter has good operability and high identification accuracy.

References

1. Yan Z, Parent C, Spaccapietra S, Chakraborty D (2010) A hybrid model and computing platform for spatio-semantic trajectories. In: International conference on the semantic web: research and applications, pp 60–75
2. Zhang Z (2010) Deriving trip information from GPS trajectories. East China Normal University
3. Wang X (2015) Identifying trips and activities using GPS based travel survey data. Jilin University
4. Yang F, Yao Z, Cheng Y, Ran B, Yang D (2016) Multimode trip information detection using personal trajectory data. J Intell Transp Syst 1–12
5. Wei L (2006) Traffic present situation analysis and traffic countermeasures research in Qingdao. Qingdao University of Technology
6. Yang J (2010) Research on clustering algorithm based on density. Changchun University of Technology
7. Ester M, Kriegel HP, Sander J, Xu X (2008) A density-based algorithm for discovering clusters in large spatial databases with noise, pp 226–231
8. Yang F, Z Yao, P J Jin (2015) Multi-mode trip information recognition based on wavelet transform modulus maximum algorithm by using GPS and acceleration data. Paper presented at Transportation Research Board 94th Annual Meeting
9. Xu B, Yi X (2004) The singularity detection of signal Basedon the wavelet transform. J Math (6):661–664
10. Heydt GT, Galli AW (1997) Transient power quality problems analyzed using wavelets. IEEE Trans Power Deliv12(2):908–915
11. Mitchell TM, Carbonell JG, Michalski RS (1986) Machine learning. Springer, US
12. Deng Z, M Ji (2010) Deriving rules for trip purpose identification from GPS travel survey data and land use data: a machine learning approach. In: International conference on traffic and transportation studies, pp 768–777
13. Gerber P (2014) Prediction of individual travel mode with evidential neural network model. Transp Res Rec J Transp Res Board 2399:1–8
14. Xiao G, Juan Z, Gao J (2015) Travel mode detection based on neural networks and particle swarm optimization. Information 6:522–535
15. Zhang H, Han Z, Li C (2002) Support vector machine. Comput Sci 29(12):135–137
16. Ding S, Qi B, Tan H (2011) An overview on theory and algorithm of support vector machines. J Univ Electron Sci Technol Chin 40(01):2–10
17. Liang Y, Reyes ML, Lee JD (2007) Real-time detection of driver cognitive distraction using support vector machines. Intell Transp Syst IEEE Trans 8(2):340–350
18. Bolbol A, Cheng T, Tsapakis I et al (2012) Inferring hybrid transportation modes from sparse GPS data using a moving window SVM classification. Comput Environ Urban Syst 36(6):526–537
19. Guo L, Ge PS, Zhang MH et al (2012) Pedestrian detection for intelligent transportation systems combining AdaBoost algorithm and support vector machine. Expert Syst Appl 39(4):4274–7286

20. Feng T, Timmermans HJP (2013) Transportation mode recognition using GPS and accelerometer data. Transp Res Part C Emerg Technol 37(3):118–130
21. Lin S, Tian F, Lu Y (2000) Bayesian networks construction and their applications in data mining. Comput Sci 27(10):69–72
22. Cheng J, Greiner R (2001) Learning Bayesian belief network classifiers: algorithms and system. Springer, Heidelberg
23. Han L, Wu S, Wang Z (2005) Bayesian belief network. Comput Knowl Technol Acad Exchange 5(21):5867
24. Liu Y, Zhang J (1993) Gradient descent method. J East Chin Inst Technol (2):12–16
25. Li X (2013) Using random forest for classification and regression. Chin J Appl Entomol 50(4):1190–1197
26. Surhone LM, Tennoe MT, Henssonow SF et al (2010) Random For 45(1):5–32
27. Hu Z (2004) A study of the best theoretical value of Gini coefficient and its concise calculation formula. Econ Res (9):60–69
28. Li Y, Zhang F, Lin Y (2000) Research on geographic information system in intelligent transportation systems. Chin J Highway Transp 13(3):97–100
29. Xu P (2007) Geography information system transportation. China Communication Press, Beijing
30. Cheng Y, Qin X, Jin J et al (2011) An exploratory shockwave approach to estimating queue length using probe trajectories. J Intell Transp Syst 16(1):12–23

Chapter 4
Mobile Phone Sensor Data Collection and Analysis

Previous studies on the application of cell phone data for traffic travel feature analysis have focused on the application of cell phone signaling data for macro travel feature extraction, i.e., the mobile operator records the cell phone user's location information of the base station cell (cell) when a signaling event (location update, call and send and receive SMS, etc.) occurs to analyze the macro travel features of traffic travelers. Cell phone sensor data have advantages in the extraction of micro travel features of individual travel chains. Traffic information based on cell phone sensor data is obtained by collecting sensor data during the whole process of travel by means of a cell phone sensor data collection app. By uploading the collected sensor data to the cloud database, the traffic data analyst can extract individual traffic travel information and laws. The current cell phone sensor data include cell phone GPS data (travel time, user location, travel velocity, etc.), accelerometer data (three-axis acceleration data), cell phone service base station data, Wi-Fi and other sensor data.

There are obvious differences in the data feature patterns of individual travel sensor data under different traffic modes, collection frequencies, traffic states and other travel conditions, which have a key influence on the identification and parameter extraction of travel chain information. Therefore, the data features must be thoroughly analyzed under different conditions to develop a more accurate and efficient algorithm for extracting individual travel feature information, which can provide a solid foundation for the application of a new generation traffic travel survey technology based on cell phone sensor data.

This chapter focuses on the development of a multisource cell phone sensor data collection app and the analysis of basic data features. Section 4.1 introduces the development method and functions of the cell phone sensor data collection app, which can realize the collection of sensor data, such as GPS data, acceleration data and service base stations. Section 4.2 introduces the establishment of a network database management system to classify and store the data uploaded by the respondents. Section 4.3 introduces the protection means, such as data encryption, used to prevent the personal information of travelers from being leaked and to ensure data security. Section 4.4 investigates five aspects: positioning accuracy and quality, travel

© Tongji University Press 2022   95
F. Yang and Z. Yao, *Travel Behavior Characteristics Analysis Technology Based on Mobile Phone Location Data*, https://doi.org/10.1007/978-981-16-8008-3_4

time dwell characteristics and OD characteristics, individual travel mobile trajectory point density characteristics, individual travel instantaneous velocity data characteristics, and acceleration data characteristics. According to the travel habits of Chinese residents who travel by means of a combination of multiple modes of transportation, a multi-scenario combination travel experiment is conducted under different combinations of transportation modes, sampling frequencies and traffic states. On this basis, the regular analysis and difference analysis of cell phone sensor data features under different scenario combinations are analyzed to achieve accurate extraction of key travel information, such as travel OD, transportation mode, transportation mode transfer time and location, in the complete travel chain of residents and to obtain refined travel characteristics of residents.

4.1 Data Collection App Development

With the rapid development of smartphone technology, smartphones have integrated a variety of smart sensors, such as GPS sensors, gravity sensors, acceleration sensors, magnetic field sensors, gyroscopes and Wi-Fi sensors. Researchers can extract effective information from some of the sensor data to conduct research and analysis of user travel behavior characteristics. In this study, a multisource cell phone sensor data collection app is developed for sensor data collection in the process of individual travel. The development platform is an Android cell phone system platform, and a backend network database management system is developed to store and manage the cell phone sensor data uploaded by users in a classified manner.

Travelers open the app before the trip, and the app automatically records the time, latitude, longitude and other data from each sensor. At the end of the trip, the traveler can use cellular mobile data or Wi-Fi to upload the sensor data collected during the trip to the database management system, completing the process of collecting and uploading travel information for the day. The development of this app is the basis for subsequent data analysis and practical applications.

4.1.1 Function Description

The app uses an Android cell phone system platform to obtain cell phone sensor data, and its detailed functions include the following.

1. Sensor Data Acquisition Function

Based on each sensor module in the cell phone, the app can collect a series of cell phone sensor data, including cell phone GPS data and motion acceleration, in real time. The GPS module in the cell phone sensor can continuously collect date, time, latitude and longitude, satellite number, and horizontal positioning accuracy

Table 4.1 Data content collected by the multisource cell phone sensor data collection app

User Name	Date	Time	Positioning Type	Longitude
13	2016/6/28	8:10:08	61	30.70778
13	2016/6/28	8:10:09	61	30.70812
Latitude	**Elevation**	**Velocity**	**Positioning accuracy**	**Number of satellites**
104.0589	388.8	3.75	18	9
104.0586	413.8	0	16	9
X-axis acceleration	**Y-axis acceleration**	**Z-axis acceleration**	**X-axis gyroscope**	**Y-axis gyroscope**
1.72	5.09	8.27	−0.0073	0.00122
1.10	5.43	8.64	−0.0085	0.00156
Z-axis gyroscope	**Direction of travel**	**MNC**	**LAC**	**CID**
0.00244	62.95	0	33,040	31,543
0.00259	65.48	0	33,040	31,543

(HDOP); the cell phone accelerometer can continuously record three-axis acceleration data in the three direction planes of the cell phone during motion. When the cell phone GPS satellite signal is blocked (such as by tall buildings or when traveling indoors or underground), the app can supplement the cell phone user in a timely manner. The current service base station cell number, Wi-Fi access and other sensor data can reflect individual travel location information. Table 4.1 shows the data content collected by the multisource cell phone sensor data collection app.

2. Personal User Information Customization Collection Function

The app can also collect the basic personal information of travelers, including name, nickname, gender, ID number, birthday, telephone number, education, and personal income. With the promotion of sensor data collection technology, more personal attributes and family attributes can be added to the basic personal information to complete the information registered in the resident travel questionnaire. When registering an account to collect information, users need to fill in only a basic user ID in the information collection interface, which can be a cell phone number or a mailbox. The rest of the user information is filled in according to the actual situation, and the specific content of the collected information can be adjusted and modified according to the actual travel situation and experimental needs of the user, which will not affect the use of the registered account. The operation interface of each functional module of the app is clear and intuitive, and the functional partition of the data collection interface is obvious and easy to identify.

3. Extensibility Features

The system hardware and software meet the requirements of the openness standard and satisfy the requirements for a future increase in system hardware nodes, expansion of database capacity, and enhancement of system software functions. When the user's needs change, the app module can be extended to meet the new needs of the user. Currently developed extensions include releasing surrounding cell phone data collection projects, which can release information about the name of data collection projects around the cell phone user, survey content and remuneration according to their geographical location, making it easier for investigators to participate in the collection project.

4.1.2   Operation Interface

The app is developed based on the Java environment. The app is a full Chinese environment, with stable and reliable operation, that is easy to operate and has high testing accuracy. The APP settings can be changed according to actual survey data needs. Moreover, the interface environment is user-friendly and fully considers the user's habits. The interface consists of the following five components.

1.   APP Startup Interface

Users can refresh the trip count information according to this interface and can start, pause or terminate the program at any time according to the actual investigation needs and user wishes. The refresh button can eliminate system faults and ensure the normal use of app functions, as well as the reliability of system operation.

2.   Data Collection Interface

The "Start" and "Stop" buttons are used to control the start time and stop time of data collection so users can easily collect travel data at the specified time, and the visualization page can display and record sensor data in real time to infer respondents' travel characteristics, including latitude, longitude, velocity, acceleration, and positioning accuracy. Respondents can verify the operation of the program through the visualization page data.

3.   Data Synchronization Interface

The "Start Upload" and "Stop Upload" buttons are used to upload the personal travel data collected via cell phone to the website database management center in a timely manner to achieve the storage and backup of survey data. In addition, the interface can display data upload progress information so that users can understand whether the data upload is normal.

4.   User Information Setting Interface

The name, gender, age, collection interval and other information settings can be adjusted by the user before starting the travel survey so that the user can collect data and the administrator can classify and filter the data in database operations.

5. Extensibility Function Interface

An information column about cell phone sensor data collection projects near the user's geographic location is provided, including the project name and payment. The user can choose one of the projects to participate in, and the project publisher will provide the user the specified payment after data collection is completed.

Figure 4.1 shows the various interfaces of the app.

(a) app startup interface (b) data collection interface

(c) data collection interface (d) user information setting interface (e) extensibility function interface

Fig. 4.1 Mobile phone sensor data collection app interface

4.2   Database Construction and Management

A network database is a system based on a backend (remote) database with certain frontend (local computer) programs to complete data storage, query and other operations through a browser. A large amount of real-time, dynamic cell phone sensor data can be obtained via the above collection app. For data storage, query management must use the network database as a powerful tool. A web-based data management system is a very effective way to classify and manage cell phone sensor data.

Once the system is established, the investigator can upload the collected sensor data to the database system. The Oracle database system is designed with SaaS architecture specifications to meet the needs of customized data collection (such as time, content, frequency customization) and multiuser parallel services. User data are stored in partitions by "project name + personal code". Data collected from different projects are stored in partitions, and extensive indexes are established based on users' personal information (name, age, cell phone number, survey time period, etc.), and data buffering mechanisms are established to ensure efficient storage and reading of data.

Furthermore, the database management system provides a data download function. Governments and enterprises can query and export data through classification attributes, and the exported data automatically generate an Excel data table. As shown in Fig. 4.2 in the data management interface, data can be accurately queried and downloaded by selecting the user's personal information index, providing massive data support for the subsequent traffic behavior analysis based on cell phone sensor data and travel information extraction work, such as travel chain acquisition.

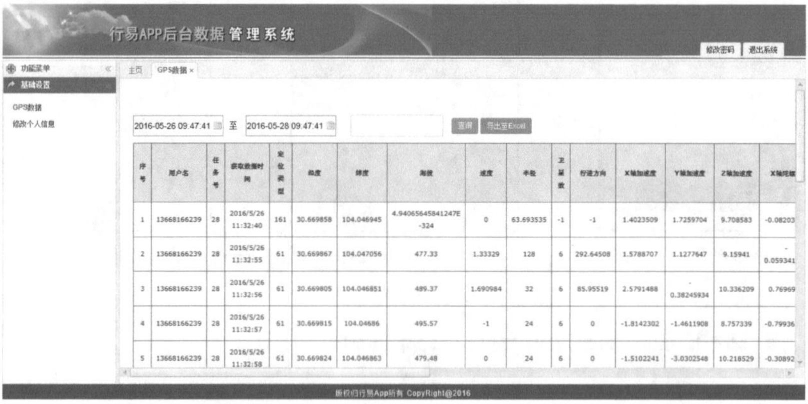

Fig. 4.2 Network database management system interface

4.3 Privacy and Data Security

The personal privacy of cell phone sensor data can be circumvented by encrypting the cell phone number and placing the processor in a mobile server room. Independent individual information is just a sample point in the massive information that is uniquely identified by a string and can be processed to obtain valuable traffic data. The specific security measures are divided into the following 3 levels of security assurance.

1. One-way encryption of cell phone number (such as C627802F5E26CEAC63A738C878F2BDA2), which uniquely identifies the phone and circumvents personal privacy issues; possible encryption algorithms to ensure that user information is not leaked include MD5, SHA, HMAC and other one-way encryption algorithms.
2. The service processing server is placed in a mobile server room to ensure the security of cell phone sensor data and data transmission velocity and physically ensure that user information is not leaked.
3. The processing result is the travel information based on the cell phone sensor movement trajectory, which is statistical data rather than individual behavior information to ensure that the user information is not leaked.

4.4 Characteristics Analysis of Mobile Phone Sensor Data

4.4.1 GPS Data Accuracy and Quality

Sources of error in the GPS module's positioning include three aspects: first, the satellite measurement error, which can be further divided into satellite clock error, ephemeris error, additional delay error in the ionosphere, additional delay error in the troposphere, multipath effect error, and the noise of the receiver itself; second, the satellite's orbital position error (for ordinary C/A code navigation GPS receivers, the horizontal positioning error is generally approximately 10 m, but the error can be less than 5 m); third, the difference in positioning error due to the difference in the selection of GPS chips because the accuracy of GPS positioning also depends on the accuracy of the selected GPS chips and the sensitivity of the GPS antennas. With the development and innovation of GPS positioning technology, the positioning accuracy of the GPS chip and the sensitivity of the antenna have been greatly improved, making it possible to obtain accurate positioning even in densely populated areas with tall buildings in large cities.

The price of GPS smartphones on the market varies, so the quality of GPS chips selected for each phone varies: low-quality GPS chips will negatively affect the quality of the collected data. In addition, due to the impact of tall buildings, underpass tunnels, overpasses and other environments, GPS track points will have missing data or data drift when actual GPS data collection is performed.

Fig. 4.3 Travel track 1 map matching result

The final matching result after logging into the network database management system to export sensor data collected by travelers, matching the complete mobile trajectory collected by each traveler to Google Maps, observing whether the matched route is consistent with the actual travel route, and the final matching results are shown in Figs. 4.3 and 4.4. As shown in Fig. 4.3, the trajectory route of the map basically matches the actual survey route, so the real-time positioning trajectory points collected by the app have a high degree of matching with the real trajectory. The matching accuracy is beneficial for the subsequent identification of transportation mode interchange points, travel OD point information and so on. In the process of matching the collected latitude and longitude data with the latitude and longitude information in Google Maps, trajectory points collected at equal time intervals are matched. Therefore, the density of the track points can also reflect the velocity of the traveller.

Furthermore, by comparing and analyzing multiple groups of travel data, it can be found that when traveling within a city, the cell phone GPS signal will be blocked by tall buildings, overpasses, etc., or underground travel signal shielding, and the GPS data will be shifted or missing, thus reducing the GPS positioning accuracy. The prerequisite for the cell phone GPS module to obtain positioning data is to ensure that there are at least three satellites to provide signals: the number of satellites that the GPS module can receive signals from decreases under overpasses or near tall buildings, which will reduce the positioning accuracy, resulting in poor signal, missing data or inaccurate positioning. Under the influence of tall buildings, the positioning data no longer match the road precisely, and it is difficult to objectively reflect the travel path and velocity characteristics of the traveller.

Fig. 4.4 Travel track 2 map matching results

4.4.2 Spatial–temporal Travel Characteristics

The origin–destination (OD) data is the core basic information for transportation system planning, design and operation management. The temporal and spatial information of each trajectory point can be obtained from the collected mobile trajectories. Spatiotemporal information is used to cluster the trajectory points collected during a day's travel to infer the movement status of travel points and analyze the travel OD.

As shown in Fig. 4.5, the planar coordinates of the spatiotemporal stay characteristics map are latitude and longitude. The Z coordinates indicate the temporal travel characteristics of the travelers, and the different colors of the travel lines represent the different moments in which the mobile trajectory points are located. The planar

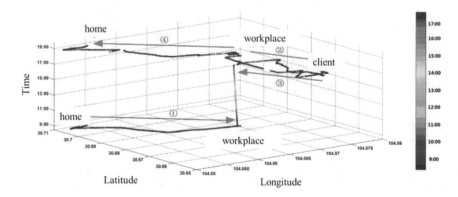

Fig. 4.5 Spatiotemporal dwell characteristics of moving track points based on sensor data

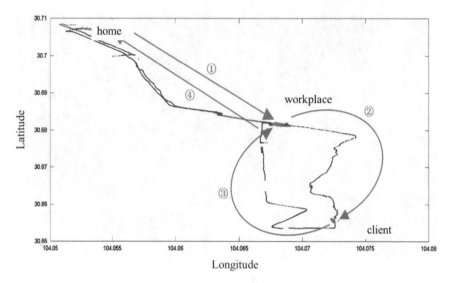

Fig. 4.6 Travel OD map of moving track points based on sensor data

X and Y coordinates indicate represent the spatial geographic locations of travelers. The length of the moving trajectory points in the Z-axis direction indicates that the traveler remains in the same location for a long time after arriving at their workplace at approximately 9:00 am. Since the Z-axis unit is 2 h per scale unit, the lag time of meeting clients from the workplace at approximately 15:30 and the lag time after returning to the workplace until the end of the afternoon are not obvious from the figure. However, by analyzing the clustering characteristics of the trajectory points and the spatiotemporal stay characteristics, we can still determine the 3 OD points and divide the travel chain into 4 travel segments.

As shown in Fig. 4.6, the travel OD features are illustrated clearly in the spatiotemporal two-dimensional diagram. Based on the clustering features and spatiotemporal stay features of the mobile trajectory points, three OD points can be identified (the red clustering points in the figure are the OD points), and the segment of the travel chain is cut into four travel segments (the blue travel trajectory points in Fig. 4.6.

4.4.3   Travel Trajectory Point Density

1.   Individual Travel Trajectory Point Density Characteristics Under Different Transportation Modes

In this section, through map matching, the collected travel sensor data of "walking-bus-walking", "walking-car-walking" and "walking-bike-walking" travel chains are imported into Google Maps for three typical combinations of transportation modes, and the path information, mode transfer point information and point density information are obtained, as shown in Figs. 4.7, 4.8 and 4.9.

As shown in Figs. 4.7, 4.8 and 4.9, the bus travel section trajectory points are sparser than the walking travel section trajectory points, the trajectory points of each section have almost equal spacing, and the trajectory points at bus stops and the interchanges between walking and bus are more dense. The trajectory of the car travel section is almost equally spaced, and there is no parking in the experimental section.

Fig. 4.7 "Walking-bus-walking" combination travel track point map

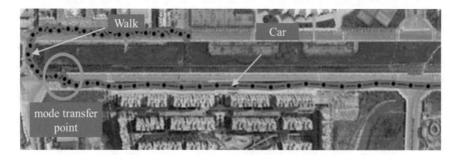

Fig. 4.8 "Walk-car-walk" combination travel trajectory point map

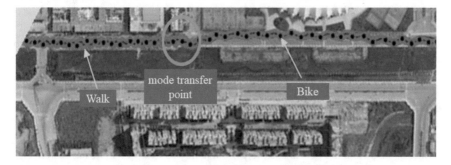

Fig. 4.9 "Walking-bike-walking" combination travel track point map

The trajectory points suddenly become dense when walking interchanges with biking, and then the bicycle trajectory characteristics are followed. The density of bicycle trajectory points is between that of bus and walking, and the distributions of trajectory points in the bicycle and walking sections are almost equally spaced. Furthermore, the density of trajectory points near the interchange point of transportation modes is the largest. Therefore, different modes of transportation have different travel trajectory change patterns, and travel velocity differences lead to different trajectory point densities. At mode interchanges, the movement velocity is 0 for a short period of time, and the travel trajectory points show a clustering phenomenon. Accurately mining the trajectory characteristics of sensor data of each traffic mode will have a decisive effect on the identification of traffic modes and traffic mode interchange points.

2.   Density Characteristics of Individual Travel Trajectory Points Under Different Sampling Frequencies

The GPS module in the cell phone sensor will continue sending information to the satellite to obtain the satellite signal during the data acquisition process and record the complete positioning information via ground information station processing. The sampling frequency of the cell phone GPS module is the frequency of acquiring data in the process of GPS information acquisition. The sampling frequency is described by the data sampling interval, i.e., the size of the interval between two adjacent GPS signals acquired by the cell phone GPS module. The smaller the data sampling interval is, the higher the sampling frequency is. For example, if the GPS sampling interval is set to 1 s, a GPS signal request is sent every 1 s, and GPS information is received and recorded; if the GPS sampling interval is set to 1 min, a GPS signal request is sent, and GPS information is received and recorded every 1 min. The collection frequency of all phone sensors is set to the same sampling frequency as the GPS module.

As shown in Fig. 4.10 (for the meanings of different color areas, refer to the attached drawings A-6), when the sampling frequency is high, such as a sampling interval of 1 s, the sensor module will acquire a large amount of data per unit time. Excessively frequent requests will greatly increase the power consumption of the phone and incur more data traffic costs. A one second sampling frequency tends to lead to a situation where an excessive amount of data is recorded, when less information is truly needed to discern travel characteristics. Therefore, data with nearly identical travel information, except for different travel times, can be considered redundant data and are not very helpful for analyzing travel characteristics.

As shown in Fig. 4.11 (for the meanings of different color areas, refer to the attached drawings A-7), a low sampling frequency reduces the power consumption of the phone during use, reduces the amount of data collected, and improves the efficiency of data storage, transmission and analysis operations. However, a sampling frequency setting that is too low will have adverse effects, especially if the amount of data is insufficient, and the accuracy of the analysis results will be reduced due to the absence of key discriminatory points. In the process of residents' travel, there is considerable information about key events, such as the point of transportation mode

Fig. 4.10 Excessive sampling frequency and data redundancy

interchange, the waiting time at intersections, and the waiting time at bus stations. All this event information is analyzed separately in the subsequent analysis of travel characteristics. These key events often do last long; for example, the total duration of mode changes between walking and motor vehicles is often approximately 10 s, but the travel characteristics are very different.

3. Density Characteristics of Individual Travel Trajectory Points Under Different Traffic Conditions

To analyze the trajectory point characteristics of individual travel movement trajectories under different traffic states, the travel survey data under three traffic states, namely, smooth, general congestion and severe congestion, were selected for analysis. Since the trajectory point characteristics of the walking and subway modes are not affected by the traffic state, the travel trajectory points of the bus travel and car

Fig. 4.11 Low sampling frequency and missing key information

Fig. 4.12 Bus trajectory point map in the clear state

travel sections are considered, and the trajectory point characteristics are obtained, as shown in Figs. 4.12, 4.13, 4.14, 4.15, 4.16 and 4.17.

Figures 4.12, 4.13 and 4.14 show that the trajectory point characteristics of buses in different traffic states are obviously different. In the completely smooth state condition, the trajectory point distribution of the bus travel section is relatively sparse and presents a more uniform distribution with no point clustering. In the general congestion condition, the trajectory point distribution is slightly dense, and some

Fig. 4.13 Bus trajectory point map under general congestion

Fig. 4.14 Bus trajectory point map under severe congestion

Fig. 4.15 Trajectory point map of a car in the clear state

Fig. 4.16 Trajectory point map of small cars under general congestion

Fig. 4.17 Trajectory point map of small cars under severe congestion

sections of the travel trajectory points are clustered. In the severe congestion state, the bus trajectory point distribution has the greatest clustering; for example, the travel section in Fig. 4.14 has 4 areas of clustered points.

It can be seen from Figs. 4.15, 4.16 and 4.17, the trajectory point characteristics of cars in different traffic states are also different. In the completely smooth state condition, the distribution of car trajectory points is relatively uniform, and there is almost no trajectory point clustering; in the general congestion condition, slight congestion causes some car trajectory point clustering. As shown in Fig. 4.16, one trajectory point clustering area occurs in each road section on average, while other trajectory points are uniformly distributed. In the severe congestion condition, the distribution of car trajectory points is the densest, and the number of clustering areas increases; for example, three areas of clustering points appear on one road section in Fig. 4.17.

4.4.4 Travel Speed Characteristics

The cell phone sensor does not record velocity data directly during the travel process; the information is calculated based on the latitude and longitude data recorded by the sensor and the cell phone sampling frequency. The developed data collection app obtains the average velocity between two points by calculating the linear distance between two sampling time points and dividing it by the sampling interval time. When the sampling interval time is sufficiently small (e.g., the sampling frequency is 1 per second), the average velocity is approximately equal to the instantaneous velocity of the sampling time point.

The instantaneous velocity values can vary substantially for different traffic mode trips and different traffic states; the fluctuation and variation in the instantaneous velocity at different sampling frequencies can also have different characteristics. Therefore, the data characteristics and variability must be analyzed.

1.   Characteristics of Instantaneous Velocity Data Under Different Transportation
     Modes

To analyze the instantaneous velocity characteristics under different traffic mode combination scenarios, four typical traffic mode combinations (i.e., "walking-bike-walking", "walking-subway-walking", "walking-bus-walking", and "walking-car-walking") are selected, with a sampling interval of 1 s. The respective instantaneous velocity–time line graphs were plotted and compared, and the results are shown in Fig. 4.18. The first and last sections show walking travel data, with velocity that are usually stable between 0 and 8 km/h, while the middle section is for bicycle travel data, with velocity that are usually stable between 10 and 20 km/h. The subway travel section is missing a GPS signal because no satellite signal is available when traveling underground, so the velocity data is missing and shown as 0. The instantaneous velocity changes abruptly when changing between walking and subway travel. The average velocity of a bus is approximately 25 km/h, and the velocity fluctuates constantly between 0 and 30 km/h. In the transition from walking to bus travel, individuals must wait at the bus stop, so there will be a period of time when the velocity is 0. The average velocity of the car is approximately 40 km/h. Because the car only stops at an intersection for red lights, the number of stops is less than that of the bus, and the fluctuation frequency is significantly lower than that of bus trips.

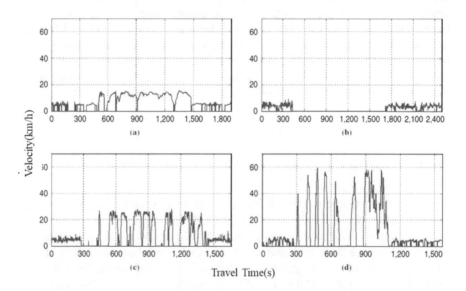

Fig. 4.18 Individual travel instantaneous velocity–time line graph for different transportation modes

According to the above analysis, the following rules can be summarized: the velocity changes abruptly near transportation mode interchange point; different transportation modes have different velocity data characteristics. Occasionally, there are temporary stopping points in the velocity curve. For the walking or bicycle mode, stops are generally due to temporary obstructions or waiting at intersections for red lights; for motorized bus and car travel, stopping is generally due to traffic congestion or waiting for traffic lights at intersections. For bus travel, temporary stops are made near each bus stop to allow passengers to get on and off, and the bus travels at a lower velocity than the car under the same traffic conditions. Buses travel at lower velocity and make more stops than small cars. Therefore, the actual velocity of the bus fluctuates less than that of the small car but fluctuates more frequently. These features will provide the basis for subsequent modeling to identify traffic interchange points and distinguish different traffic modes.

2. Characteristics of Instantaneous Velocity Data Under Different Sampling Frequencies

To analyze the characteristics of the instantaneous velocity of individual trips with different sampling frequencies, the complete travel chain of "walking-bus-walking" with a sampling interval of 1 s is used as a typical sample, and the sparsification process of equal time length is performed. The selected time lengths for sparsification are 5, 10, 20, 30, 60 and 120 s, and the resulting velocity data characteristics are shown in Fig. 4.19.

Figure 4.19 shows that as the sampling interval increases, the time interval between data points increases. When the sampling time interval is 30 s or less, the velocity characteristics and fluctuation trend are more obvious; when the interval increases to 60 s, the data volume greatly decreases and the fluctuation trend weakens; and when the interval is 120 s, the data volume decreases substantially and the velocity characteristics no longer noticeably change, making it difficult to distinguish different traffic modes.

3. Characteristics of Instantaneous Velocity Data Under Different Traffic Conditions

Under different traffic states, the traffic flow will have different characteristics. Significant differences in traffic velocity and traffic continuity occur between smooth and congested states. Because motor vehicle traffic flows are spatially separated from pedestrian and nonmotorized bicycle trips on sidewalks, the road traffic state does not have as much of an influence on the instantaneous velocity of the walking and bike modes (in addition, the traffic state has no influence on subway operation). This section focuses on the instantaneous velocity and variability of buses and cars in three traffic states, smooth, general congestion and severe congestion, with data collected with a 1-s sampling interval.

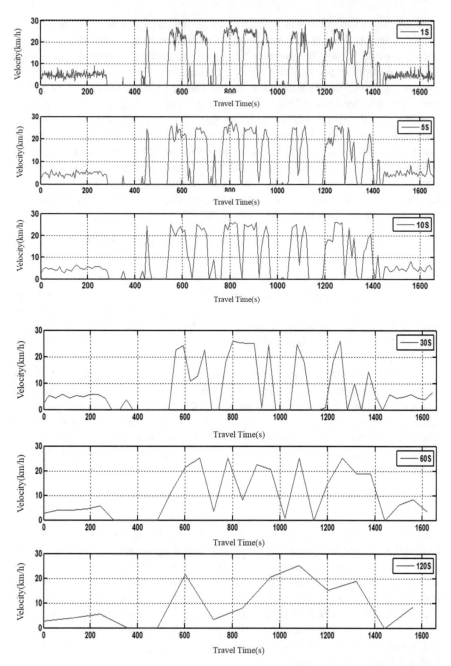

Fig. 4.19 Individual travel instantaneous velocity–time line graph at different sampling frequencies

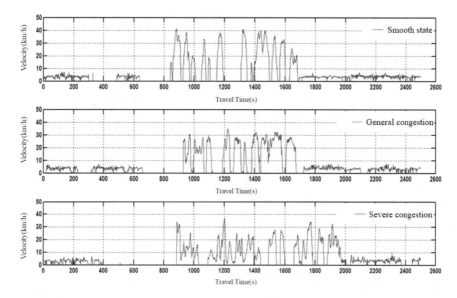

Fig. 4.20 Characteristics of instantaneous velocity data for different traffic states of the "walking-bus-walking" combination

(1) Characteristics of Instantaneous Velocity Data in the Three States for the "Walking-Bus-Walking" Combination

In this section, three sets of individual travel data for the "walking-bus-walking" combination under different traffic conditions are selected, and the velocity–time line is shown in Fig. 4.20.

As shown in Fig. 4.20, the characteristics of the "walking-bus-walking" combination are as follows: after walking for a while from the starting point, a person must wait at a bus stop, board the bus, exit the bus and continue walking to the end. Bus trips stop at bus stops and red lights. The bus velocity distribution in different traffic states has obvious variability. In the smooth state, the peak instantaneous velocity of the bus travel section reaches approximately 40 km/h; under general congestion, the peak velocity of the bus travel section is approximately 30 km/h; under severe congestion, the peak velocity is approximately 30 km/h but is generally approximately 20 km/h. Moreover, the instantaneous velocity is as low as 10 km/h in some periods.

(2) Characteristics of Instantaneous Velocity Data in Three Traffic States for the "Walking-Car-Walking" Combination

To analyze the instantaneous velocity characteristics of the "walking-car-walking" travel mode combination under different traffic conditions, three sets of individual travel survey data are selected in this section, and the velocity–time line is shown in Fig. 4.21.

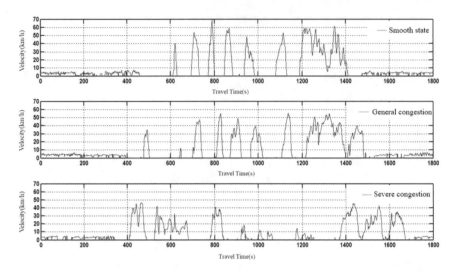

Fig. 4.21 Characteristics of instantaneous velocity data for different traffic states for the "walking-car-walking" combination

As shown in Fig. 4.21, the "walking-car-walking" combination consists of three stages: walking from the starting point to the taxi or taxi point, getting into the car and taking the car to the parking position near the destination, and then walking to the end point. According to the velocity–time diagram of car trips in the clear state, cars mainly stop at red lights at intersections. Therefore, cars generally stop less than buses. In the smooth state, the peak velocity of small cars is approximately 70 km/h; in the general congestion state, the peak velocity is approximately 55 km/h, with some segments having velocity values below 40 km/h; in the severe congestion state, the peak velocity is also approximately 45 km/h, and most of the velocity values are below 20 km/h.

(3) Comparison of Instantaneous Velocity Characteristics Between Buses and Cars in Three Traffic States

Since the instantaneous velocity data characteristics of buses and cars are relatively similar, to further explore the characteristics and differences between the instantaneous velocity of the two travel modes, this section combines mathematical and statistical methods to analyze the interval distributions of buses and cars in different traffic states, as shown in Figs. 4.22 and 4.23.

According to Fig. 4.22, in the smooth state, the bus velocity is mainly distributed in the 10–40 km/h interval, and the average distribution ratio of each subinterval is approximately 30%; in the general congestion state, the distribution ratio of the 20–30 km/h interval is the largest, reaching 56.68%, followed by the 10–20 km/h interval, reaching 27%; in the severe congestion state, the two intervals of 0–10 km/h and 10–20 km/h are the largest, at approximately 40%, followed by the 20–30 km/h interval at 18.82%.

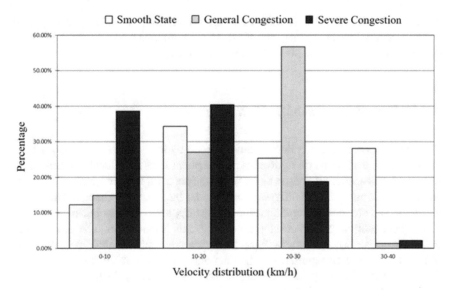

Fig. 4.22 Bus velocity distribution under different traffic conditions

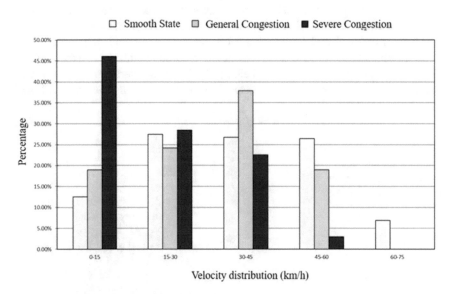

Fig. 4.23 Speed distribution of small cars under different traffic conditions

According to Fig. 4.23, the velocity of small cars in the smooth state is mainly evenly distributed in the 15–50 km/h interval, with approximately 26% in each subinterval; However, 6.86% is in the 60–70 km/h interval. In the general congestion state, the velocity is mainly distributed in the 30–45 km/h interval (37.92%), followed by the 15–30 km/h interval (24.16%). In severe congestion, the velocity distribution shows a trend from high to low, with the 0–15 km/h interval being the most common (46%), followed by the 15–30 km/h interval (28.44%).

4.4.5 Travel Acceleration Characteristics

The acceleration data are collected and recorded by the acceleration sensor of the mobile phone. The signal strength of the mobile phone does not affect the acceleration sensor. The acceleration sensor has low power consumption and high sensitivity, and the acceleration data of different travel modes have specific characteristics. The accelerometer obtains acceleration values along three axes of the mobile phone plane: X-axis acceleration, Y-axis acceleration and Z-axis acceleration. The acceleration values comprise two components: gravity acceleration and acceleration caused by different modes of transportation. The data characteristics under other conditions can be extracted by analyzing the acceleration data to identify the traffic modes and traffic transfer points (Fig. 4.24).

Due to constant changes in the mobile phone plane during the actual collection process, the directions of the X-, Y- and Z-axes are uncertain. To effectively mine the change characteristics of acceleration, this study first synthesized the three-axis acceleration and analyzed the data characteristics of different traffic modes according to the combined acceleration. Suppose for any recorded point, the X-axis acceleration is x, the unit is m/s^2 (the same below), the Y-axis acceleration is y, the Z-axis acceleration is z, and the combined acceleration is a. The relation is defined as follows:

$$a = \sqrt{x^2 + y^2 + z^2} \tag{4.1}$$

Fig. 4.24 Mobile phone acceleration sensor and its measurement of the three axis directions

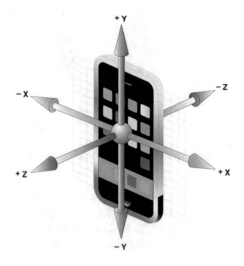

1. Data Characteristics of Instantaneous Acceleration Under Different Modes of Transportation

Figures 4.25, 4.26, 4.27 and 4.28 show the fluctuation characteristics of three-axis acceleration with a sampling interval of 1 s (obtained from the calculation of mobile

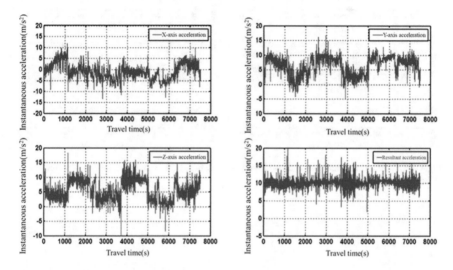

Fig. 4.25 "Walking-bicycle-walking" combination of triaxial acceleration and combined acceleration-time line graph

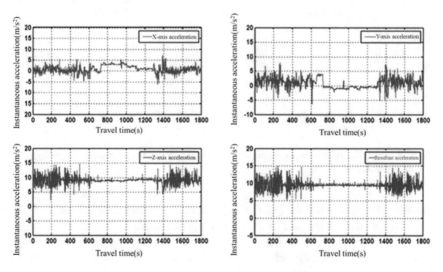

Fig. 4.26 "Pedestrian-subway-pedestrian" combination of the three axes and combined acceleration-time line diagram

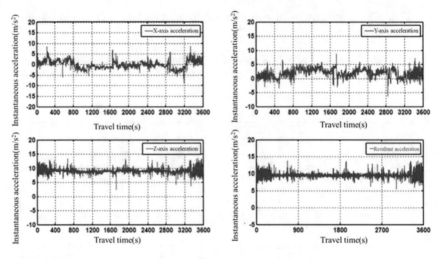

Fig. 4.27 Triaxial acceleration and combined acceleration-time line diagram of the "walking-bus-walking" combination

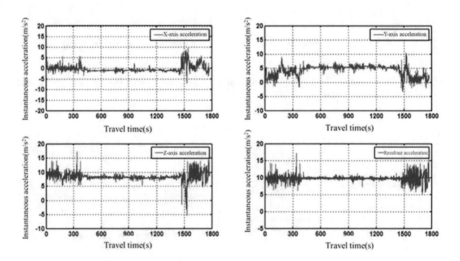

Fig. 4.28 Triaxial acceleration and combined acceleration-time line diagram of the "walking-car-walking" combination

phone acceleration sensor data) under different traffic mode combinations. The fluctuation range of acceleration of walking and bicycles is greater than that of buses or cars mainly because the experimenter's body continues moving, and the amplitude is larger when walking and cycling. Simultaneously, body movement is minor when standing or sitting in a vehicle.

As shown in Fig. 4.25, for "walking-bike-walking", the acceleration along the X-axis, Y-axis and Z-axis fluctuates substantially. The acceleration of the X-axis fluctuates from -5 to 5 m/s^2, that of the Y-axis fluctuates regularly from 0 m/s^2 to 10 m/s^2, and that of the Z-axis fluctuates regularly from 0 to 10 m/s^2, and the periodicity of the fluctuation is more obvious. The combined acceleration is much more stable and fluctuates slightly around 9.8 m/s^2 (gravitational acceleration is 9.8 m/s^2). This is because when walking and cycling, the body is moving more, and the three plane directions of the mobile phone are continually changing. Therefore, three-axis acceleration cannot accurately reflect the acceleration in all directions.

In contrast, although the combined acceleration introduces gravitational acceleration, which does not reflect the travel characteristics of the mode, it comprehensively considers the acceleration components of the travel mode acceleration along the three axes, so it can reflect the acceleration characteristics of the traffic mode as a relatively ideal index. From the perspective of the data fluctuation amplitude, the noise in the travel acceleration data along each axis can be offset when synthesized into the combined acceleration. However, no significant difference is observed between the combined accelerations of walking and cycling in terms of numerical values. The two transportation modes are difficult to distinguish by means of the value and fluctuation of acceleration; further characteristic analysis of the acceleration data is needed.

As shown in Fig. 4.26, in the combination of walking and subway, the fluctuation in acceleration along the X-axis, Y-axis and Z-axis of the subway travel segment are significantly smaller than those of the walking travel segment. The acceleration along the X-axis fluctuates from -5 to 5 m/s^2, that of the Y-axis fluctuates from -5 to 5 m/s^2, and that of the Z-axis fluctuates between 5 and 15 m/s^2. In terms of the triaxial acceleration, there is a significant difference between the subway trip and the walking trip. The fluctuation range of the combined acceleration in the walking travel segment is also extensive, similar to that of the triaxial acceleration. In terms of combined acceleration, the subway travel section is relatively stable, and most of the subway runs smoothly, with fluctuation at only a few points because subway travel will not encounter intersections or traffic jams. Acceleration and braking are performed only at fixed station locations, so the frequency of acceleration fluctuation is minimal.

As shown in Fig. 4.27, for the "walking-bus-walking" combination, the acceleration along the X-axis and Y-axis in the bus travel segment and the walking travel segment constantly fluctuates, with no significant difference in the extent of the upward and downward fluctuations. The acceleration along the X-axis fluctuates from -5 to 5 m/s^2, that along the Y-axis fluctuates from -5 to 5 m/s^2, and that along the Z-axis fluctuates between 5 and 15 m/s^2. No significant difference in the numerical value is observed between the X-axis and Y-axis acceleration in bus travel and walking travel, so they are not easy to distinguish. For Z-axis acceleration and combined acceleration, the walking trip segment's fluctuation range is more extensive, similar to the fluctuation range of three-axis acceleration when walking. However, the bus travel segment's fluctuation frequency is greater than that of the subway, and the fluctuation range is also more extensive. Buses are slow and need to accelerate and brake at bus stops. Compared with the subway trip in Fig. 4.26, the number of acceleration fluctuations is much greater.

As shown in Fig. 4.28, for the "walking-car-walking" combination, both the triaxial acceleration and the combined acceleration can be used to distinguish the travel modes of walking and car travel. The acceleration along the X-axis fluctuates from -5 to 5 m/s^2, that along the Y-axis fluctuates from 0 to 5 m/s^2, and that along the Z-axis fluctuates between 0 and 15 m/s^2. In terms of triaxial acceleration and combined acceleration, the fluctuation range of the walking segment is larger. The fluctuation frequency of the car travel section is less than that of bus travel, the fluctuation range is smaller, and the car runs smoothly. This is because cars do not need to stop at bus stops, so they do not accelerate and brake as often.

In summary, Figs. 4.25, 4.26, 4.27 and 4.28 show that most of the X-axis and Y-axis acceleration is generated by travel mode acceleration and fluctuates in an extensive range around approximately 0. Z-axis acceleration and combined acceleration are generated by the acceleration of gravity and the acceleration of the travel mode. When walking and riding a bicycle, the Z-axis acceleration and combined acceleration fluctuate considerably. When taking a subway, bus or car, the acceleration fluctuates less. Nevertheless, all fluctuate around approximately 9.8 (the gravitational acceleration is approximately 9.8). Although the combined acceleration accounts for gravitational acceleration, which does not reflect the travel characteristics, it comprehensively considers the acceleration components of the travel mode along the three axes to create an ideal index that can thoroughly reflect the travel mode's acceleration characteristics.

Some noise acceleration data on the triaxial component are offset when synthesized into the combined acceleration, so the quality of the combined acceleration data is relatively high.

2. Data Characteristics of Instantaneous Acceleration at Different Sampling Frequencies

Similar to the analysis of instantaneous velocity data at different sampling frequencies, when the sampling frequency is high, there are more redundant data in the acceleration data, which affects the data processing efficiency. When the sampling frequency is low, it is easy to miss critical discrimination points, which will cause difficulties in the analysis of the subsequent acceleration data.

The complete travel chain of "walking-bus-walking" with a sampling time interval of 1 s is taken as a typical example to perform sparse processing with time intervals of 5, 10, 20, 30, 60, and 120 s. The processing results are shown in Figs. 4.29, 4.30, 4.31, 4.32, 4.33 and 4.34.

Figures 4.29, 4.30, 4.31, 4.32, 4.33 and 4.34 show that when the sampling time interval is less than 10 s, the acceleration variation characteristics and fluctuation trend are relatively prominent. When it is increased to 30 s, the amount of data is significantly reduced, and fluctuation is weakened. Moreover, when the interval time is 60 or 120 s, the data volume decreases significantly. In this case, the acceleration characteristics have no noticeable changes, so it is not easy to distinguish travel modes.

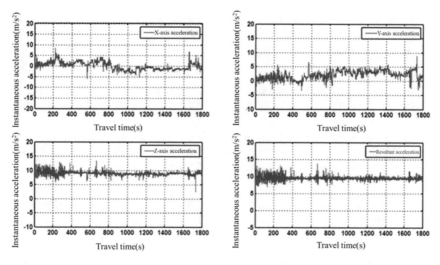

Fig. 4.29 Triaxial acceleration and combined acceleration-time line plot at 1 s sampling interval

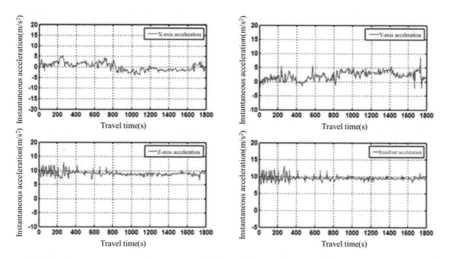

Fig. 4.30 Triaxial acceleration and combined acceleration-time line plot at 5 s sampling interval

3. Data Characteristics of Instantaneous Combined Acceleration Under Different Traffic Conditions

The same method is used to analyze instantaneous velocity. This section examines the characteristics and differences of buses' and cars' combined acceleration under three traffic conditions: unblocked, regular traffic, and severe traffic.

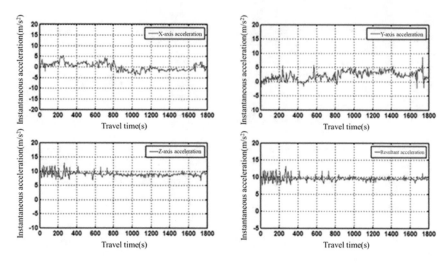

Fig. 4.31   Triaxial acceleration and combined acceleration-time line plot at 10 s sampling interval

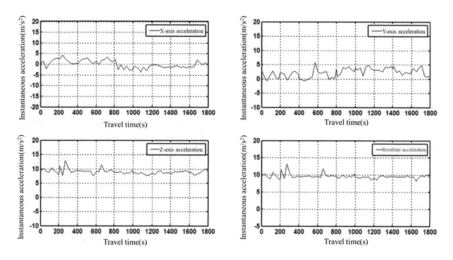

Fig. 4.32   Triaxial acceleration and combined acceleration-time line plot at 30 s sampling interval

(1)   "Walking-Bus-Walking" Instantaneous Combined Acceleration in Three States

In this section, three groups of instantaneous combined acceleration data of the "walking-bus-walking" combination under different traffic conditions are selected. The graphed acceleration-time polyline is shown in Fig. 4.35. The journey consists of three stages: walking from the starting point and waiting at the bus stop, getting on the bus and riding the bus, getting off the bus and walking to the destination. In the unblocked state, the bus's maneuverability is the best, and the acceleration fluctuates considerably. In the general congestion state, bus mobility worsens, and

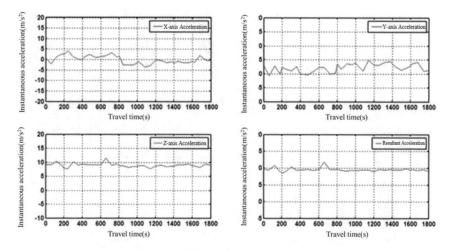

Fig. 4.33 Triaxial acceleration and combined acceleration-time line plot at 60 s sampling interval

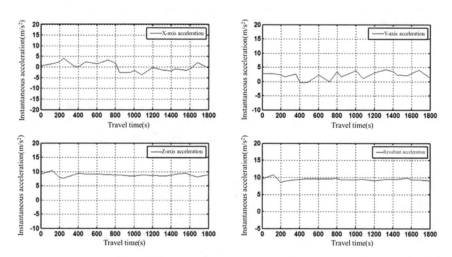

Fig. 4.34 Triaxial acceleration and combined acceleration-time line plot at 120 s sampling interval

the acceleration fluctuation decreases. Finally, in severe congestion, the change in acceleration is the smallest, and the instability of acceleration in some sections is more significant. The fluctuation range of acceleration in some areas is almost the same as that of walking.

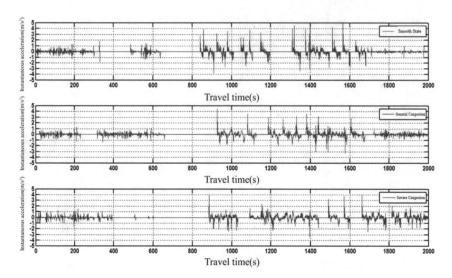

Fig. 4.35 Data characteristics of instantaneous combined acceleration in different traffic states for the "walking-bus-walking" combination

(2)   "Walking-Car-Walking" Instantaneous Combined Acceleration Data in Three States

As shown in Fig. 4.36, the combined travel process of "walking-car-walking" includes three stages: walking to the bus or taxi-point, taking a car to reach the destination, and walking to the destination. It would help if you stopped when driving

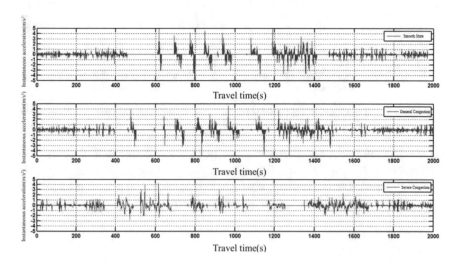

Fig. 4.36 Data characteristics of instantaneous combined acceleration in different traffic states of "pedestrian-car-pedestrian"

at a red light at an intersection. Therefore, the number of stops for car trips is less than that for bus trips, as reflected in the acceleration graph; that is, the number of fluctuations is less. Under conditions of free-flowing traffic, regular congestion and severe congestion, the acceleration fluctuation amplitude of cars is similar to that of buses and will not be described here.

(3) Comparison of Instantaneous Velocity Characteristics Between Buses and Cars in Three Traffic States

Since the instantaneous velocity data characteristics of buses and cars are relatively similar, to further explore the characteristics and differences between the instantaneous velocity of the two travel modes, this section combines mathematical and statistical methods to analyze the interval distributions of buses and cars in different traffic states, as shown in Figs. 4.37 and 4.38.

According to Fig. 4.37, it can be concluded that under three different traffic conditions, the distribution of bus combined acceleration is characterized as "high in the middle and low at both ends". Among them, the acceleration distribution ratio between -1 and 1 m/s^2 is the largest, with each interval being about 40%, followed by the interval between -2 and -1 m/s^2 and between 1 and 2 m/s^2. It can be seen that the more traffic jams, the more frequent starting and braking, the smaller the range of acceleration fluctuations, and the more concentrated in the interval from -2 to 2 m/s^2.

According to Fig. 4.38, it can be concluded that in three different traffic states, the combined acceleration distribution of cars presents the feature of "high in the middle and low at both ends". In the state of severe congestion, the distribution ratio from -1 to 1 m/s^2 is the largest, which accumulates to 80%. The second is the general congestion state, the two interval cumulative distribution rate reached about 60%; Finally, the smooth state, the cumulative distribution rate of the two intervals reached about 50%.

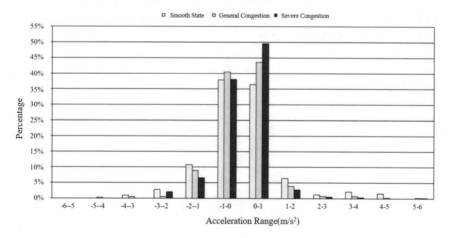

Fig. 4.37 Bus velocity distribution under different traffic conditions

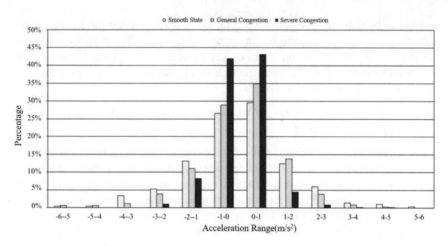

Fig. 4.38 Velocity distribution of small cars under different traffic conditions

4.5   Summary

This chapter introduces a multisource mobile phone sensor data acquisition app, the construction of the network database management system, data security measures and mobile phone sensor data characteristics and difference analysis. The analysis of data characteristics and differences is the core content of this chapter. This chapter studies the positioning accuracy and quality of the GPS module, the spatiotemporal stay characteristics and OD characteristics of individual travel modes, the density characteristics of personal journey tracking points, individual travel instantaneous velocity and the features of acceleration data. The mobile phone sensor data obtained under different traffic modes, sampling frequencies and traffic conditions are combined to analyze the data under different conditions. This paper lays a solid data foundation for accurately extracting critical travel information from individual travel chains and obtaining refined travel characteristics.

Chapter 5
'Pedestrian-Traffic Flow-Communication' Integrated Simulation Platform Construction

With the development of the social economy, smartphone penetration rates are increasing daily, providing an excellent opportunity for the analysis of individual transportation activities based on big mobile data. Mobile phone navigation software has gradually penetrated people's lives because of its convenience, and people have steadily become familiar with satellite positioning, such as mobile phone GPS. Moreover, traffic surveys based on mobile phone sensor data have become possible. Due to its high positioning accuracy, low survey costs and many advantages, this approach can compensate for the shortcomings of traditional resident travel surveys.

Furthermore, the rapid evolution and application of the 3G/4G-LTE new generation of mobile communication technology has injected new power into big mobile data. The upgrading of the mobile communication network from traditional 2G will enhance mobile phone location tracking. The frequency with which mobile phone location changes can be identified will increase, and mobile phone tracking surveys will improve. For example, when surfing the Internet, app software such as WeChat and QQ interacts with the mobile communication network every few minutes and automatically reports the phone's location. With the rapid increase in users of such social software, mobile phone location data will undoubtedly revolutionize the development of transportation networks, which will further improve the accuracy of traffic data of mobile users and improve the reliability of traffic big data environments. The 3G/4G LTE new generation of mobile communication technology can significantly improve mobile phone positioning capability, as reflected in the following points:

1. The transmission rate of the communication network was enhanced from the traditional 2G GSM network rate of 9.6 k/s to 144 k–2 m/s, which significantly reduces signal transmission delay and provides adequate communication broadband support for applications that require higher frequency, such as WeChat, QQ, and video signal data.
2. "Soft handover" or "softer handover" is used to effectively eliminate the "ping-pong handover" in the hard handover and increase the reliability and stability of the handover location.

© Tongji University Press 2022
F. Yang and Z. Yao, *Travel Behavior Characteristics Analysis Technology Based on Mobile Phone Location Data*, https://doi.org/10.1007/978-981-16-8008-3_5

3. The mobile phone location mobility management ability is enhanced, and the mobile phone location area's updated information can be more accurate.

These favorable factors are expected to further improve the quality and accuracy of traffic data collection.

However, due to personal privacy, network security, and other sensitive issues, it has been challenging to obtain a large amount of sufficient actual mobile phone location data to evaluate the application of such technology in various traffic, communication, and other environments throughout the year. In reality, road traffic conditions, travel environments and a variety of uncontrollable situations are very complicated; sometimes some specific traffic state may occur but cannot be generated, which greatly hinders the particular environment of mobile phone positioning technology applied in traffic data collection research. Furthermore, the vast majority of existing research results are based on the use of actual mobile phone communication data or sensor data to extract traffic parameters. Nevertheless, due to the lack of accurate traffic information as a comparative check, researchers often find it challenging to accurately evaluate research results.

Therefore, the integrated pedestrian and traffic flow simulation platform built in this chapter and the 3G/4G-LTE new-generation mobile communication simulation platform can be used to study the controllable system state parameters, comprehensively evaluate the impact of various system parameters and influencing factors, and provide a deeper and comprehensive understanding of the applicability of the technical method.

Section 5.1 presents a brief overview of the simulation platform's construction framework. Section 5.2 provides a summary of the current traffic simulation software and the basic principles of traffic simulation and builds a traffic simulation module. Section 5.3 simulates the user's wireless communication events, outlines the signal propagation principle of mobile communication, and develops a mobile communication simulation module based on the 3G/4G-LTE new-generation mobile communication mechanism. In Sect. 5.4, the individual multimode travel data derived from the traffic simulation module are perturbed, mobile phone sensor data are simulated, and the simulation data are preliminarily evaluated. Finally, Sect. 5.5 summarizes the content of this chapter.

5.1 Framework of the Simulation Platform

This chapter's integrated simulation platform, which integrates traffic flow simulation and 3G/4G-LTE new-generation mobile communication simulation, consists of three main modules, namely, the traffic simulation module, mobile communication network signaling event simulation module and mobile phone app sensor data extraction module.

The architecture of the mobile phone communication simulation platform is shown in Fig. 5.1. Traffic simulation module: according to the configured accu-

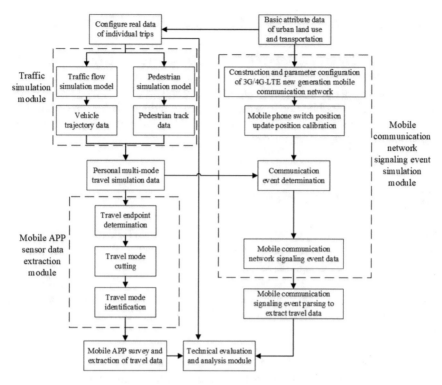

Fig. 5.1 The architecture and process of the simulation platform that integrates pedestrian and traffic flow simulation and new-generation mobile communication simulation

rate data of travel individuals, the existing traffic simulation model and software are used to transform the data into a multimode travel simulation trajectory data set, which provides a data source for mobile phone sensor data simulation and serves as a reference comparison for subsequent communication data and mobile phone sensor data mining traffic flow information. Phone app sensor data simulation module: through the survey statistics of the probability distribution function of a GPS positioning error, many travel simulation models export trajectory-based data scrambling data sets so that the scrambling mobile sensor data and actual data present similar wave characteristics to ensure the similarity of the simulation data and fundamental data. Communication simulation module: a mobile communication network signaling event simulation module is built according to the actual 3G/4G-LTE new-generation mobile communication network, base station layout and parameter configuration. The statistical probability function of individual communication activities and behavior events is obtained through investigation and statistical analysis to generate individual communication behaviors. The communication events are loaded into the mobile communication simulation module to generate the dataset of mobile communication network signaling events.

5.2   Traffic Environment and Individual Travel Simulation

The network and road conditions of traffic simulation are configured according to the natural traffic environment. Traffic flow data similar to those in the natural environment are generated through the simulation method and serve as the data source for the simulation data of mobile phone sensors and provide the basis for the comparative analysis of the subsequent extraction of travel characteristics using mobile phone communication data and sensor data.

5.2.1   Traffic Environment Design

The VISSIM simulation software developed by the German PTV Company is widely used domestically and abroad for traffic simulation. VISSIM simulation can provide an intuitive, detailed image of vehicle, road, intersection, and lights over time, to accurately reproduce the traffic network running status. It provides a feasible new idea and method to identify intersections and even whole city traffic organization schemes under the complicated mixed traffic conditions, with good operability and practicability [1].

In the construction of the simulation platform in this book, the traffic road network diagram is imported into VISSIM software to build a VISSIM traffic network that is consistent with the actual situation. Then, the traffic volume is loaded according to the traffic volume of each road obtained from the traffic survey so that the traffic environment in VISSIM is consistent with reality. VISSIM can generate visual traffic operation status online and output various statistical data offline, such as vehicle velocity, travel time, and queue length. In the subsequent wireless communication simulation module, when individual communication events occur, the timely position data derived from VISSIM serve as the data source for interacting with the base station, thereby generating the communication signaling data set. In the mobile phone sensor simulation module, the location trajectory and velocity data derived from VISSIM serve as the ideal data source for mobile phone sensor simulation.

5.2.2   Individual Travel Module Construction and Simulation

Traffic simulation is performed according to the real trajectory of individual travel, and the simulation data attributes include travel OD, travel mode, travel path, travel time, dwell time and so on. MATLAB software is used to obtain the walking travel trajectory data (travel position coordinates, velocity, etc.). The simulation results reflect pedestrians' interaction and the interaction between pedestrians and the surrounding environmental obstacles. Pedestrian movement can be realized by searching for the target cost function with the minimum perception. The corresponding vehicle travel trajectory data are generated with VISSIM traffic simulation software, which conforms to the characteristics of traffic flow motion parameters. The trajectory data of individual multimode travel simulations are obtained by connecting the two types of trajectory data.

This section takes Chengdu, Sichuan Province, as an example to build a simulation model. Chengdu has a built-up area of 604.1 square kilometers, a permanent resident population of 14.428 million, a dense traffic network and a complex traffic situation. Chengdu is a city with typical traffic travel characteristics and preferences. This book conducts simulation for the second ring road, where the traffic volume is most concentrated. The traffic flow and signal timing of the main road section are considered. The VISSIM road network of the Chengdu second ring road is set up and vehicles are loaded, as shown in Fig. 5.2. According to the survey results, the traffic flow and signal timing in the simulated road network are loaded to reflect the actual traffic state as accurately as possible.

The individual travel information derived through VISSIM includes the following: personal number, simulation time, target path number, current path number, vehicle type, X coordinate, Y coordinate, and velocity (m/s). Since the coordinate information exported by VISSIM is based on VISSIM coordinates, it can be linearly converted into longitude and latitude. The conversion equation used in this book is as follows:

$$Longitude = 1.0156 * 10^{-5} * Vissim\_X + 103.8766;$$
$$Latitude = 9.03546 * 10^{-6} * Vissim\_Y + 30.49497; \tag{5.1}$$

Finally, the exported individual multimode travel data is shown in Table 5.1.

 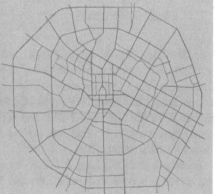

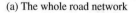

(a) The whole road network   (b) Vehicle load

Fig. 5.2 VISSIM network of Chengdu second ring road

Table 5.1 Individual multimode travel trajectory data

Individual number	Simulation time	Destination path number	Current path number	Vehicle type	Longitude	Latitude	Velocity (m/s)
10	1622	10,231	229	300	104.0587	30.67668	11.94
10	1623	10,231	229	300	104.0588	30.67663	12.17
10	1624	10,231	229	300	104.0589	30.67657	12.4
10	1625	10,231	229	300	104.059	30.67652	12.42
10	1626	10,231	229	300	104.0591	30.67646	12.18

5.3   Wireless Communication Simulation

The purpose of the mobile communication network signaling event simulation module is to generate the mobile communication network signaling event data set according to the simulated individual communication behavior and travel trajectory under the 3G/4G-LTE new-generation mobile communication environment. This approach provides a conditionally controllable data source for subsequent research on traffic flow parameter extraction, such as individual travel trajectory mining based on communication signaling data. In addition, although previous studies have made many achievements in this area based on actual communication data, the accuracy of their research results is challenging to assess due to the lack of real traffic information as a reference. The module's simulation data can be used as the data source to evaluate the accuracy and technical feasibility of the extraction algorithm.

5.3.1   Wireless Communication Events Description and Simulation

Traditional wireless communication events include mobile phone calls and short messages. According to a report on China's broadband penetration released by the China Broadband Development Alliance, by the second quarter of 2016, Chinese mobile broadband (mostly 3G and 4G) user penetration rate reached 63.8%. With the popularization of 3G/4G-LTE new-generation mobile communication network technology, wireless Internet access has become an essential component of wireless communication events [2].

Based on mobile phone bill data from a large number of mobile phone user volunteers, this book conducts statistical analyses to fit the probability distribution of various communication events and determines the statistical probability function of communication activities. The individual multimode travel simulation trajectory data generated in the traffic simulation module are loaded according to the probability and statistical functions of communication activities and behavior events to configure the parameters of the travel trajectory to simulate individuals' mobile phone use

Table 5.2 Representation values of different communication events

State	Leave unused	Received messages	Call behavior	3G/4G-LTE
Representative value	0	1	2	3

behavior during travel, including calls, short messages, WeChat, QQ and other events that trigger 3G/4G Internet access and other information interaction events with the communication base station. To statistically analyze and generate the communication events in a convenient manner, different communication events are represented by numbers, as shown in Table 5.2.

1. Call Event Simulation
(1) Call Time Simulation

Call duration refers to the time elapsed between the beginning and end of a call. Statistical analysis of 96 consecutive hours of telecom call data in a city, including more than 3 million pieces of information about users' Internet access, such as number of calls, call time, call date, call duration and base station information, indicates that the call duration of users in different periods is significantly different [3]. The average call length peaked between 22:00 and 23:00 and then fell sharply, reaching its lowest point between 6:00 and 7:00 in the morning. By contrast, during working hours, the average call time is relatively stable until 8:00 PM, as shown in Fig. 5.3.

The least-squares method was used to fit the curve; the fitting effect of the 5-degree Gaussian polynomial was the best. The values of each function are shown in Table 5.3. The root mean square error (RMSE) was 1.534, and the coefficient of determination (R-square) was 0.9983. Figure 5.4 shows the fitting effect diagram, in which the black points are scattered points on the actual curve, and the blue line is the fit curve, indicating that the fitting effect is good. According to the individual travel time, the user's average call time can be calculated, and a normal distribution can be generated.

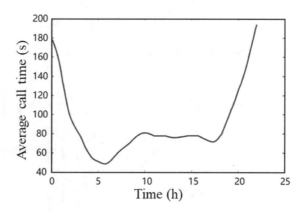

Fig. 5.3 The average call time of users in different time periods

Table 5.3 Fifth-degree Gaussian polynomial fitting function value

Parameter	Value	95% confidence interval
a1	210.4	(149.3, 271.4)
b1	23.04	(21.02, 25.07)
c1	3.117	(−0.6226, 6.856)
a2	4.805e + 16	(−2.387e + 19, 2.397e + 19)
b2	−195.3	(−3114, 2723)
c2	33.95	(−218.1, 286)
a3	27.81	(−45.13, 100.7)
b3	19.42	(18.91, 19.93)
c3	1.421	(0.4425, 2.399)
a4	70.02	(54.81, 85.24)
b4	10.07	(9.323, 10.81)
c4	4.061	(3.217, 4.906)
a5	66.11	(44.33, 87.88)
b5	16.02	(15.21, 16.84)
c5	3.55	(1.878, 5.222)

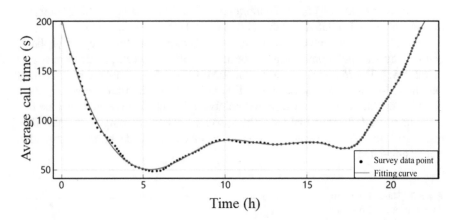

Fig. 5.4 Quintic Gaussian polynomial fitting effect

$$f(x) = a1 * \exp\left(-\left(\frac{x - b1}{c1}\right)^2\right) + a2 * \exp\left(-\left(\frac{x - b2}{c2}\right)^2\right)$$

$$+ a3 * \exp\left(-\left(\frac{x - b3}{c3}\right)^2\right) + a4 * \exp\left(-\left(\frac{x - b4}{c4}\right)^2\right)$$

$$+ a5 * \exp\left(-\left(\frac{x - b5}{c5}\right)^2\right) \tag{5.2}$$

Table 5.4 User call interval time distribution table

Interval between call events	Frequency	Cumulative percentage (%)
0–10 min	166	40.29
10–20 min	39	49.76
20–30 min	21	54.85
30–40 min	22	60.19
40–50 min	15	63.83
50–60 min (1 h)	15	67.48
60–70 min	10	69.90
70–80 min	11	72.57
80–90 min	17	76.70
90–100 min	13	79.85
100–110 min	7	81.55
110–120 min (2 h)	9	83.74
120–130 min	6	85.19
130–140 min	3	85.92
140–150 min	9	88.11
150–160 min	4	89.08
160–170 min	3	89.81
170–180 min (3 h)	4	90.78
Other	38	100.00

(2) Call Interval Simulation

Call interval time refers to the time span between two calls. The smaller the value is, the more frequently the user makes calls. The probability that the interval between two consecutive calls in the surveyed population is 0–30 min is 54.85%, and the probability decreases with increasing interval time because mobile phone calls have a certain continuity. Simultaneously, the probability that the time between calls is greater than three hours is approximately 10%, and this percentage is likely to increase as 3G/4G-LTE networks become more widely available in the future. See Table 5.4 for the detailed distribution of the call interval time. The roulette selection method is used to randomly generate the call interval time, that is, to generate a random number X within 0–1. The median of the call event time interval in this segment is taken as the median of the normal distribution, and a random number is generated from the normal distribution as the call time interval [4].

2. Message Event Simulation

The short message interval time refers to the time span between sending or receiving consecutive messages. The time interval between message sending and receiving of

Table 5.5 Distribution of message receiving and sending intervals

Interval	Frequency	Cumulative percentage (%)
0:00:00	2279	18.38
0:05:00	4733	56.57
0:10:00	724	62.41
0:15:00	329	65.06
0:20:00	208	66.74
0:25:00	195	68.31
0:30:00	141	69.45
0:35:00	132	70.51
0:40:00	97	71.30
0:45:00	124	72.30
0:50:00	101	73.11
0:55:00	77	73.73
1:00:00	66	74.27
2:00:00	652	79.53
3:00:00	401	82.76
4:00:00	253	84.80
5:00:00	211	86.50
10:00:00	594	91.30
15:00:00	437	94.82
20:00:00	254	96.87
24:00:00	155	98.12
else	233	100.00

the surveyed population in a single month showed apparent regularity. The interval of less than one hour accounts for almost 75% of the data because there is considerable continuity between sending and receiving a message. Mobile phone users often use messages to chat or discuss things with their relatives and friends, and multiple messages are often sent within a short period. The detailed interval data are shown in Table 5.5. The roulette selection algorithm is used to calculate the interval, and a random number drawn from a normal distribution with the department's median value as the mean is taken as the SMS receiving and sending interval.

3. 3G/4G-LTE Wireless Network Communication Events

With the popularization of Internet apps, a large number of terminals send heartbeat packets periodically. The heartbeat packet was initially used as a secure backup mechanism for servers. To prevent servers from crashing, unique ports and lines were used to send short messages between servers periodically [5]. Internet applications on mobile phones have borrowed this mechanism. Android native apps, QQ, Weibo and WeChat all use a heartbeat mechanism in which the terminal sends a short message to the application server at regular intervals. Various mobile phone apps have different heartbeat cycles. For example, the heartbeat cycle of the old version of QQ was 30 s, that of the latest version of QQ is 180 s, and that of WeChat is 300 s, which has shown a gradually increasing trend. Due to the variety of apps on different users' mobile phones, the rules are difficult to investigate and statistical analysis is complicated. This book takes a random number between 50 to 400 s as the heartbeat cycle to simulate 3G/4G-LTE mobile communication events.

5.3.2 Mobile Communication Signal Propagation Simulation

The study of electromagnetic wave propagation characteristics is an essential basis for the coverage of mobile communication networks. The station signal strength is used to judge switching. The signal from the base station is transmitted to the phone in the form of electromagnetic waves. Because the transmission path between the base station and the mobile phone is relatively complex, from simple line-of-sight propagation to encountering various types of terrain obstructions (tall buildings), coupled with the diversification of the electromagnetic wave propagation mechanism (reflection, diffraction and scattering), the radio signal strength varies substantially. This is the root cause of mobile switching stochastic volatility. The change characteristics of signal intensity at a specific position point of the range signal transmitter can be understood based on the study of electromagnetic wave propagation characteristics to provide a basis for the analysis of mobile phone switching [6].

1.   Weakness Theory of Electromagnetic Wave Propagation

The modeling of electromagnetic wave propagation is a difficult task in wireless mobile communication systems. Propagation model research traditionally focuses on predicting the average received signal field intensity at a certain distance from a long-distance signal transmitter and the variation of the signal field intensity near a specific position. The attenuation of a propagating electromagnetic wave is manifested in two main forms: path loss and small-scale fading. Path loss is the attenuation of signal intensity caused by distance between the receiver and transmitter. Small-scale fading refers to the phenomenon that a mobile phone may experience rapid fluctuations in the instantaneously received signal strength when moving within a small range. Therefore, two kinds of electromagnetic wave propagation models are developed based on distance variation: ① a large-scale path loss propagation model, which is used to describe the variation of the signal field intensity over a long distance (several hundred meters or several kilometers) between the transmitter and receiver; ② a small-scale fading model, which describes the rapid fluctuation of the received field intensity within a short distance (several wavelengths) or a short time (seconds).

(1)   Large-Scale Path Loss

The primary problem in wireless communication system design is how to calculate the intensity or the median value of a received signal under the given conditions to design other parameters and indicators of the system. The given conditions include transmitter antenna height, position, working frequency, receiving antenna height, and the distance between transceivers. This scenario is the prediction problem of radio wave propagation path loss. The median signal refers to the median value of long interval [7].

Because of the mobile environment's complexity and variability, the median of the received signal is difficult to calculate accurately. The approach of wireless communication engineering is to determine the relationship between the propagation loss (or the received signal field strength) and the distance, frequency and antenna height under various topographical features based on a large number of field intensity tests and statistical analysis of the data. Various charts and calculation equations of propagation characteristics are presented, and a prediction model of wave propagation is established; this simple method can be used to predict the median of the received signal.

In mobile communication, many prediction models of radio wave propagation, which are summarized according to the measured field intensity data in various terrains and have features that are suitable for different situations, have been established. A combination of analysis and experimentation is used to produce most propagation models. The experimental method is based on fitting an appropriate curve or analytical equation to a series of measurement data. The advantage of this approach is that all the transmission factors, including known and unknown, are considered based on actual measurements.

(2)   Small-Scale Fading (Fast Fading)

Small-scale fading affects the signal level's dynamic fading range on the receiver's antenna in the local region. The random fluctuation of the signal intensity caused by small-scale fading significantly influences the switching judgment. Small-scale fading refers to the fact that the amplitude, phase or multipath delay of a radio signal changes rapidly over a short time or a short propagation distance, so the influence of large-scale path loss can be ignored. This fading is caused by the interference of the same transmission signal, which travels along two or more paths and arrives at the receiver with a slight difference. These waves are called multipath waves. The receiver antenna synthesizes these waves into a signal that changes dramatically in amplitude and phase, depending on the multipath wave's strength, relative travel time, and bandwidth of the transmitted signal.

In general, in central urban areas with tall buildings, because the mobile antenna height is lower than the surrounding buildings, no single line-of-sight propagation path exists from the mobile station to the base station, which leads to fading. Even if a line-of-sight propagation path exists, multipath propagation remains due to the ground's reflection from surrounding buildings. Thus, incident waves arrive from different directions and have different propagation delays.

2. 3G/4G-LTE Data Analysis of Signaling Events in the New-Generation Mobile Communication Network

The new-generation communication system architecture adopts a flat network design mainly composed of the core network and access network. The flattening of the web makes each entity function more centralized. The system signaling and processing are simplified, and the change in the network architecture alters the cellular mobile communication system change from circuit-switched mode to packet-switched mode. In this case, the user device's location function is also enhanced and provides a means to support the implementation of the location service. Compared to the GSM network, the new-generation communication system provides more abundant, accurate and convenient location information services.

Signaling resolution process of a new generation of mobile communication: mobility management of the cellular network includes tracking area (TA) updating in the idle state and handover in the connected state. The former periodically triggers position update signaling, while the latter triggers the corresponding soft handover process via the handover algorithm specified by the system [8]. Frequent interactive signaling in a cellular mobile communication system includes location information, such as location marker TAI and cell marker. The extraction of such signaling information can provide good support for tracking individual travel in the transportation industry. For example, Fig. 5.5 shows the soft handover process of the new mobile communication network generation.

New generation of mobile communication network signaling event information extraction: the popularity of smartphone terminals makes QQ, WeChat, Weibo and

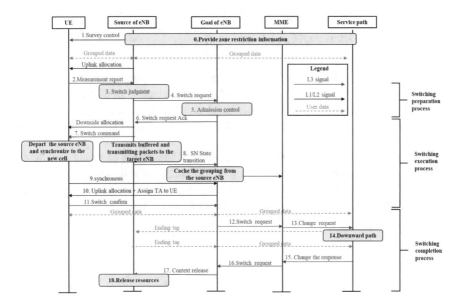

Fig. 5.5 Soft handover process in the new-generation mobile communication network

other mobile phone apps widely used. The multimedia functions of such software, such as text, voice, and video, provide people a convenient and fast way to communicate and produce a large amount of network positioning information. From the perspective of traffic, this kind of software generates massive signaling information and frequently exposes users' location information, which is beneficial in the study of individual traffic travel trajectories. For the 3G/4G-LTE new-generation communication system, the wireless interface protocol includes the control plane and user plane, which can be analyzed in the user interface of the relevant application protocols. Network packet analysis tools such as Wireshark are used to extract location-related data streams generated by GPS, OTDOA and other positioning technologies.

3.   Wireless channel model description

In communication, the channel model measures the signal strength of mobile phone users in a complex ground environment. In this book, SNR is used as a criterion to assess the service base station ID of each sampling location, and repeated experiments are conducted [9].

$$SNR = \frac{P_T - PL - S}{N} \tag{5.3}$$

where, P_T is the base station power, PL is the path loss, S is shadow fading, and N is noise power. The Winner II model, which is shown below, is used for path loss.

$$PL = \begin{cases} 21.5\log_{10}(d) + 44.2 + 20\log_{10}\left(\frac{f_c}{5}\right), & 10\,\text{m} < d < d_{BP} \\ 40\log_{10}(d) + 10.5 - 18.5\log_{10}(h_{BS}) \\ -18.5\log_{10}(h_{MS}) + 1.5\log_{10}\left(\frac{f_c}{5}\right), & 10\,\text{m} < d < d_{BP} \end{cases} \tag{5.4}$$

In the equation, d, h_{BS}, and h_{MS} are the distance between the base station sampling points, base station height (typical value 32 m), and sampling point height (typical value 1.5 m), respectively, and the unit is m. f_c is the carrier frequency, and the unit is GHz. d_{BP} is the distance of the break point $d_{BP} = 4h_{BS}h_{MS}f_c/c$. c is the velocity of light.

Network construction is performed according to the layout planning principles of the 3G/4G-LTE new-generation mobile communication network, and the corresponding communication network parameters are configured simultaneously. The main parameter configuration content is shown in Table 5.6.

Table 5.6 Wireless communication simulation parameter configuration

Parameter	Value
Bandwidth/MHz	4
Total transmitting power of base station/dB	43
Path loss/dB	130.19 + 37.6lgR (R is the distance between the base station and the mobile phone, km)
Logarithmic shadow fading standard deviation/dB	4
Thermal noise/dB. Hz^{-1}	−174
User noise index/dB	7
Type of antenna	Omnidirectional antenna
Base station cable loss/dB	2
Hysteresis margin/dB	8
Frequency/GHz	2.5

5.3.3   A Case Study of Wireless Communication Simulation

The communication simulation model of Chengdu is reconstructed based on the traffic simulation. Figure 5.6 shows the simulation results of the signal coverage distribution law of the GSM base station in Chengdu at a specific time. The hollow circle represents the base station's location, the red line is the road contour, and the different colors around the base station represent the coverage range of signals from

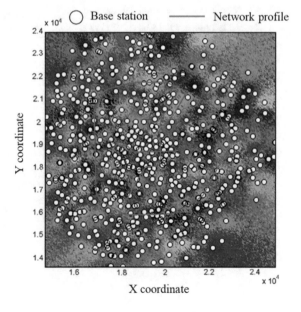

Fig. 5.6 Simulation results of the coverage range of the communication base station in Chengdu

different base stations. Because the signal intensity emitted by the base station is affected by environmental factors (architecture, weather, etc.) and decreases with distance, the coverage range of the base station at different times also changes dynamically [10].

① The coverage range of each base station cell is diverse (depending on the transmitting power of the base station and the attenuation law of signal propagation). ② Due to the disturbance of signal propagation, the base station cell boundary is irregular and changes dynamically with time. ③ Multiple cell base stations may cover the same location. The simulation results of the coverage distribution of these base stations closely match the actual communication environment. The mobile phone wireless communication simulation data platform proposed in this book can effectively compensate for the defects of existing mobile phone positioning models. For example, the simulation data platform completely avoids the assumptions that the base station cell covers the same area and presents the ideal hexagonal cell shape, thereby increasing the authenticity of the simulation results. The above analysis indicates that the mobile phone wireless communication simulation platform proposed in this book can effectively simulate the natural communication environment and provide necessary support for further extraction of mobile phone positioning data.

Table 5.7 shows a sample of simulated mobile phone communication data. The first 8 represent individual trip data generated by the traffic simulation module, and CT is the simulated wireless communication event data. CI is the sequence of interactive base stations obtained through mobile communication simulation; CLng and CLag are the base station latitude and longitude. If no communication event occurs, the CT column is 0, and the last three columns are also 0.

Table 5.7 Mobile communication data sample

II	ST	TL	LN	VT	Lng	Lat	S[m/s]	CT	CI	CLng	CLat
10	2353	11	183	300	104.0729	30.62992	11.89	0	0	0	0
10	2354	11	183	300	104.0729	30.62981	12.12	0	0	0	0
10	2355	11	183	300	104.0729	30.6297	12.35	0	0	0	0
10	2356	11	183	300	104.0729	30.62958	12.47	3	428	104.0721	30.62736
10	2357	11	183	300	104.0729	30.62947	12.23	3	428	104.0721	30.62736
10	2358	11	183	300	104.0729	30.62936	12	3	427	104.0721	30.62738
10	2359	11	183	300	104.0729	30.62926	11.77	3	428	104.0721	30.62736

Note II—Individual ID; ST—Simulation Time; TL—Target Link; Ln—Link Number; VT—Vehicle Type; Lng—Longitude; Lat—Latitude; S[m/S]—Velocity ([m/S]); CT—Communication Type; CI—Cell ID; CLng—Cell Longitude; CLat—Cell Latitude

5.4 Mobile Phone Sensor Data Simulation

Due to sensitive issues such as personal privacy and network security, sensor data of actual travel are difficult to obtain. Moreover, artificial experimental collection requires considerable time and labor costs, especially in specific traffic environments and traffic states. It is challenging to collect mobile phone sensor data and control the experimental conditions, but simulation can substantially reduce the cost of data collection. Moreover, the system state parameters are easier to maintain, which is conducive to comprehensive evaluation of the technical method's application effect under conditions of various system parameters and influencing factors.

5.4.1 Data Disturbance Loading Method and Simulation

Because the trajectory data obtained through pedestrian simulation and traffic flow simulation are ideal trajectory points, each anchor point must be scrambled. The simulated sensor data have similar data characteristics to those of real location data.

Many domestic scholars have researched GPS positioning errors. Xu Kun established the GPS dynamic single-point positioning model using the time series method based on 2640 positioning data points collected from a road section with light traffic flow in Xi'an at a driving velocity of 60 km/h [11]. The statistical results are shown in Table 5.8.

The latitude and longitude errors are obtained by drawing random numbers from a normal distribution that meets the specified conditions. Since the GPS offset direction is closely related to the vehicle driving direction, this statistical result does not apply to the error statistics in all directions. However, the offset distance can be calculated on the basis of the statistical results. The offset distance is a scalar quantity, and the error distribution has certain universality.

Li Jun took 508,511 pieces of data collected by 1485 floating vehicles in Guangzhou all day as the research object and obtained the distribution law of the offset road error [12]. The calculation of the offset road positioning error is shown in Fig. 5.7. The average error is 12.14 m, the variance is 151.11 m^2, and the positioning error follows a P-norm distribution. The expression is as follows:

Table 5.8 Statistics of latitude and longitude errors

	Maximum absolute error/($'$)	Mean square error/($'$)	Mean absolute error/($'$)
Longitude original error sequence	0.2410	0.0802	0.0636
Latitude original error series	0.7407	0.2018	0.1500

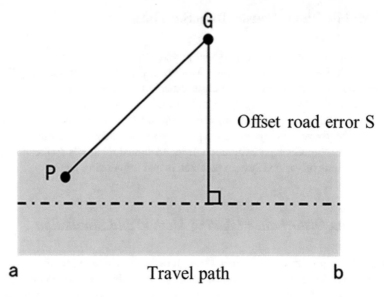

a   Travel path   b

Fig. 5.7 Positioning error

$$f(x) = \frac{p\lambda}{2\sigma\Gamma\left(\frac{1}{p}\right)} \exp\left[-\left(\lambda\frac{|x-\mu|}{\sigma}\right)^p\right] \tag{5.5}$$

where, $\Gamma(x)$ is the gamma function, $\lambda = \sqrt{\frac{\Gamma\left(\frac{3}{p}\right)}{\Gamma\left(\frac{1}{p}\right)}}$, $p = 0.94$, $\sigma = 12.19$, and the goodness of fit is 0.98. Random numbers from the P-norm distribution can be generated to obtain the deviation of the road distance from the error point.

Based on the offset distance and the positioning error from the road, combined with the vehicle's driving direction, the location of the offset GPS location point can be determined. As shown in Fig. 5.8, Point P is the real registration point, and line PQ is the vehicle's driving direction. By determining the offset distance P and the offset road error S, point G's position after disturbance can be determined.

The simulation was conducted on the basis of the above statistical results. Taking a group of data derived from VISSIM as an example, the trajectory diagram and velocity line diagram before and after the disturbance are shown in Figs. 5.9 and 5.10. In Fig. 5.9, the red points are the ideal positioning trajectory, while the blue points are the location trajectory points after disturbance. The blue simulated trajectory points deviated from the ideal position. In Fig. 5.10, the red line is the velocity line diagram derived from VISSIM, and the blue line is the velocity line diagram after disturbance. The velocity derived from VISSIM has strong regularity and stability. In contrast, the velocity after disturbance fluctuates violently and is similar to the velocity obtained from satellite positioning in reality.

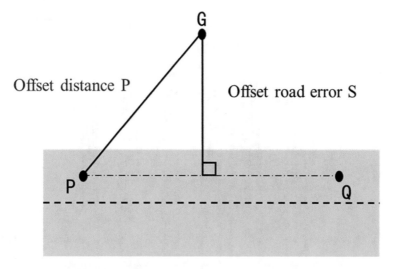

Fig. 5.8 GPS point positioning offset

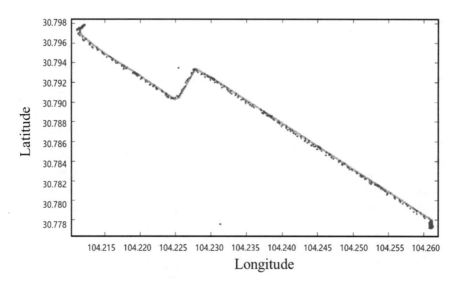

Fig. 5.9 Comparison of original and disturbed trajectories

5.4.2 A Case Study of Mobile Phone Sensor Data Simulation

A test line was designed for field collection and platform simulation for comparative evaluation to verify the difference between the data generated by simulation and the actual mobile phone sensor data. Shuhan Road—Shudu Dadao—Dongdajia—Dongda Road was selected as the test route. The route runs east—west, starting

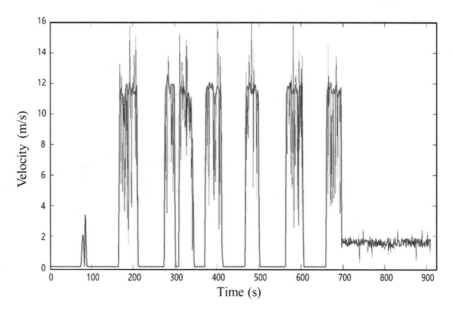

Fig. 5.10 Comparison of original and disturbed velocity

from Gelin Garden on Shuhan Road, Second Ring Road of Chengdu in the east, and
ending at Huan Trade Square on Dongda Road in the west. The route covers a total
length of approximately 10.7 km. Shuhan Road, Shudu Avenue, Dong Dajie and
Dongda Road are all major traffic arteries. Furthermore, this line includes subway
line 2, bus 43 and other bus lines, passing through the People's Park, Tianfu Square,
the Affiliated Hospital of Chengdu University of Traditional Chinese Medicine and
other areas where people gather and distribute densely.

The travel mode of the test data is walking—bus—bicycle—walking. Traffic simulation of the sample data is performed, and mobile phone sensor data are generated via
disturbance processing. Figure 5.11a, b show the trajectory of the GPS data collected

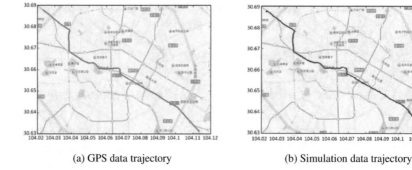

(a) GPS data trajectory   (b) Simulation data trajectory

Fig. 5.11 GPS data and simulation data trajectory

in the field and the simulated data trajectory, respectively. The collected and simulated trajectory data follow the same route on the map. Figure 5.12a shows the collected velocity data and the transportation mode adopted at that time. Figure 5.12b shows the corresponding relationship between the traffic mode and velocity generated by the simulation. The actual collected and simulated velocity data show similar fluctuation characteristics. However, due to the influence of many uncontrollable factors in the real world, the bus in the actual collection process runs at a low velocity and stops at a slightly higher frequency than it does in the simulation data.

1. Assessment of the Overall Distribution of the Simulation Data

To evaluate the sensor simulation data's authenticity, the velocity distribution of different traffic modes is statistically analyzed. The results are shown in Fig. 5.13. The blue and red lines are the velocity distribution of the actual data collected in the field and the mobile phone sensor simulation data, respectively. Different velocity

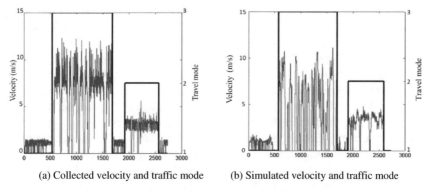

(a) Collected velocity and traffic mode   (b) Simulated velocity and traffic mode

Fig. 5.12 Field collection and simulation of GPS velocity-traffic mode broken line diagram

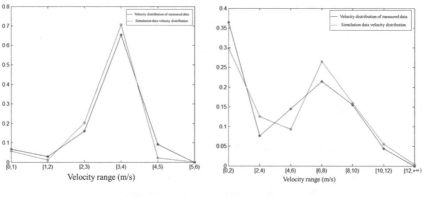

(a) Collected velocity and traffic mode   (b) Simulated velocity and traffic mode

Fig. 5.13 Distribution of the measured and simulated velocities

interval distributions refer to the proportion of the running time within a given velocity interval. Figure 5.13a shows the velocity distribution of bicycle transportation. As bicycle travel is less affected by environmental conditions than are motor vehicles, the driving environment is similar to that in the simulation, and the proportions of different velocity intervals present a relatively consistent distribution. The simulation data are in good agreement with the measured data. Figure 5.13b shows the velocity distribution of buses. Due to the strong sensitivity of bus velocity to traffic, the difference in the velocity distributions of the field data and simulation data is larger than that for the bicycle. For example, the simulation data has a higher proportion of buses with a normal running velocity of 6 m/s or greater because the real traffic environment is complex, and vehicles start and stop frequently. However, the proportion difference between the measured data and the simulation data in different velocity distribution ranges is all within 0.1, which is within the acceptable range.

2.   Evaluation of the Velocity Simulation Data

The abovementioned statistical analysis of the velocity distribution reflects the fitting degree of the simulated and real velocity data. The similarity of the two is assessed on the basis of several velocity characteristics.

Different traffic modes have different sensitivities to road conditions; for example, motor vehicles start and stop more frequently than nonmotor vehicles, which means that the velocity variation range of different modes of traffic differs substantially. Therefore, the standard deviation of velocity per unit time is an important feature to identify traffic modes. Furthermore, the difference between the average velocity and the maximum velocity per unit of time is correlation with the transportation mode. Figure 5.14 compares two kinds of velocity data with these three velocity characteristics within 60 s after each time point. Although a certain deviation in the numerical value is observed, the fluctuations show similar characteristics.

The above evaluation and test show that the trajectory data and velocity data of the mobile phone sensor obtained after the disturbance of individual multimode travel derived from the traffic simulation module are in good agreement with the collected real data. These two kinds of data are the essential characteristic quantities of mobile phone sensor data. The sensor simulation data generated in this chapter can be applied in subsequent research.

5.5   Summary

By building an integrated simulation platform of "pedestrian, traffic flow and communication", the effect of the technical method can be studied under an environment with controllable system state parameters, and the effect of different system parameters and influencing factors can be comprehensively evaluated, which is conducive to obtaining a more profound and comprehensive understanding of the applicability of the technical method. This chapter takes the Chengdu Second Ring Road as an

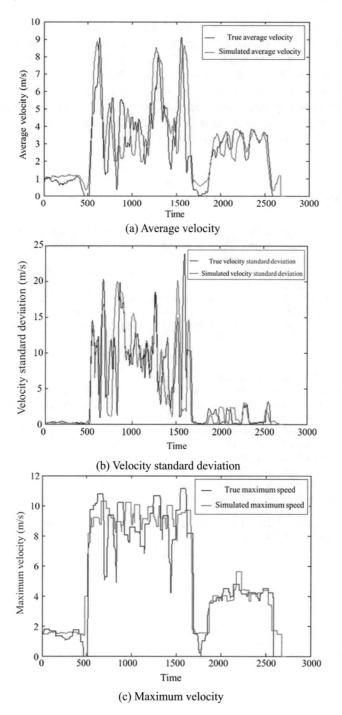

(a) Average velocity

(b) Velocity standard deviation

(c) Maximum velocity

Fig. 5.14 Comparison of different velocity characteristics of simulation data and real data

example, generates traffic trajectory data of individual travel modes through VISSIM and MATLAB software, and simulates the personal communication signaling data according to the network construction, base station layout and parameter configuration of the real 3G/4G-LTE new-generation mobile communication. Finally, the statistical analysis of GPS positioning error, load disturbance of the road traffic simulation trajectory data, simulated mobile sensor data, and real mobile sensor data showed good consistency between the real and simulated data. Thus, this paper provides a data basis for the subsequent use of simulation data.

References

1. Park B, Schneeberger JD (2003) Microscopic Simulation model calibration and validation: case study of VISSIM Simulation model for a coordinated actuated signal system. Transp Res Rec J Transp Res Board (5):185–192
2. Anonymous (2016) Broadband development alliance: China's broadband penetration rate is around 60%. TvEng 40(9):152
3. Zhang W (2013) Analysis in behavioural characteristics of telecommunication customer. University of Electronic Science and Technology of China, Beijing
4. Lipowski A, Lipowski D (2011) Roulette-wheel selection via stochastic acceptance. Phys A Stat Mech Appl 391(6):2193–2196
5. Sun Y (2014) Key words in communication industry in 2013: letter and 4G. UPS Applic 1:82–82
6. Chang L (2013) Research on radio propagation channel models and time-varying properties of mobile communication systems. Beijing University of Posts and Telecommunications, Beijing
7. Gong D, Liu C, Guo J (1998) Discussion of path loss formulas for evaluating third-generation mobile systems. Telecommun Netw Technol 4:38–41
8. Lang W, Yang D (2011) Mobility management in LTE idle state. Data Commun (6):6–9
9. Meinilä J, Kyösti P, Jämsä T et al (2008) WINNER II channel models. In: Radio technologies and concepts for IMT-Advanced. Wiley, pp 39–92
10. Fei Y (2013) Link travel speed data capture technology based on cellular handoff information: method, algorithm and evaluation. Science Press
11. Xu K, He Y, Sun C et al (2008) Pavement crack positioning based on GPS. J Chang'anUniv Nat Sci Ed 28(5):27–30
12. Li J, You Y, Deng H (2016) Distribution characteristics of FCD positioning error in urban areas. SciTechnol Eng 1:244–249

Chapter 6
Empirical Study on Trip Information Extraction Based on Mobile Phone Sensor Data

Individual travel parameters are the basis for the analysis of urban traffic demand, including the starting and ending points of each trip, the mode of transportation, and the purpose of travel. The individual travel information survey method based on mobile phone sensor data (time, GPS latitude and longitude data, velocity, three-axis acceleration) has advantages in the extraction of microtravel characteristics of individual travel chains. This approach has the characteristics of high data accuracy and low operation difficulty, which can effectively compensate for the shortcomings in traditional travel surveys. The previous chapters introduced mobile phone sensor data analysis technology, mining algorithms, application methods and conditions and proposed algorithm support for the application of mobile phone sensor data. Empirical analysis of the actual application effect is still needed.

Based on the mobile phone sensor data travel parameter extraction algorithm introduced in Chap. 3, this chapter uses the developed multisource mobile phone sensor data collection app to conduct surveys and data processing and summarizes the difficulties and technical keys of using mobile phone sensor data to extract individual travel parameters. The three travel characteristic parameters of the travel endpoint, transportation mode transfer point, and transportation mode are extracted and analyzed, and a comprehensive evaluation of the feasibility of the algorithm is conducted.

Section 6.1 designs an investigation scheme and comprehensively collects mobile phone sensor data in line with the traffic characteristics of residents in China. Section 6.2 empirically analyzes the recognition effect of the spatial clustering algorithm on the travel endpoint and discusses the feasibility and effect of travel endpoint recognition in mobile phone sensor data. Section 6.3 studies the wavelet transform modulus maximum algorithm to identify the transfer points of traffic modes and then analyzes the differences in travel data of different traffic modes and identifies travel segments. Sections 6.4 and 6.5 study the ability of the neural network algorithm to identify the travel mode and improve the recognition accuracy through a GIS map matching algorithm. Finally, Sect. 6.6 summarizes the results and analyzes the feasibility of individual travel parameter extraction.

© Tongji University Press 2022   151
F. Yang and Z. Yao, *Travel Behavior Characteristics Analysis Technology
Based on Mobile Phone Location Data*, https://doi.org/10.1007/978-981-16-8008-3_6

6.1 Experiment Design and Data Collection

Comprehensive analysis of the characteristics of mobile phone sensor data and the extraction of individual travel parameters requires data that fully reflect the travel characteristics of Chinese residents. China's transportation environment is complex, with a large population of residents, and daily travel is characterized by a large travel volume and diverse travel modes. Measured data should cover the normal travel of most residents in order to extract individual travel parameters for analysis. This part analyzes the characteristics of residents' travel, develops data collection plans for different characteristics, and conducts surveys in Chengdu.

China's urban residents' transportation trips have the following characteristics:

(1)   The average number of trips per person in most cities is less than 3/(person·d), and the average number of trips per person in a few cities exceeds 3/(person·d). Under normal circumstances, the number of trips per capita of residents decreases with increasing city size and decreases with increasing travel distance.

(2)   The purpose of travel is divided into commuting and noncommuting, each accounting for approximately half of the trips, and commuting has regular temporal and spatial characteristics.

(3)   There are various travel modes, most of which involved a combination of multiple modes.

The development of the mobile phone sensor data collection plan requires a comprehensive analysis of the processing effects of mobile phone sensor data on different traffic conditions based on the characteristics of China's residents' traffic travel and consideration of the diversity of the city's several arterial roads to analyze the characteristics of different traffic modes in Chengdu. With a population of more than 14 million permanent residents, Chengdu is an urban area with a high population density and is representative of China's complex transportation environment. This study designed 13 data collection routes. The travel plan covers common daily travel modes consisting of walking, bicycles, buses, cars and subways, such as walking-transit-transit-walking (transfer at the same station), walking-transit-walk-bus-walk (transfer between different stations), walk-bus-walk-metro-walk, walk-rent-metro-walk, walk-bike-bus-walk, etc. The sample size of each mode combination is 30 trips, and the trip plan takes into account multiple factors, such as trip purpose and traffic status.

6.1.1 Travel Plan for Different Travel Purposes

Different travel purposes impact traveler choice, travel time and distance, and the corresponding travel data characteristics are also very different. To verify the adaptability of the algorithm for the extraction of individual travel parameters, it is necessary to extract and verify travel data under the influence of various factors. This section divides the travel composition of residents according to the purpose of travel, designs a travel plan based on the traffic conditions in Chengdu, and collects specific travel data for fixed travel purposes.

According to the survey results of relevant cities, residents' travel composition is divided into commuting and noncommuting according to the purpose of travel. Commuting includes commuting to get to work and school, which accounted for 41.7% and 6.5% of the total trips, respectively. Noncommuting trips accounted for 51.9% of the total trip composition, including travel for daily life, cultural entertainment and leisure, transportation, medical treatment and business, which accounted for 25.4%, 9.8%, 9.3%, 3.2%, and 4.2%, respectively. Based on the above ratio and considering normal travel rules, such as the choice of mode, the time of occurrence corresponding to the travel purpose, and whether the length of the journey is reasonable, a travel survey plan based on different travel purposes is designed to include commuting and noncommuting travel.

The commuting route simulates the commuting behavior of residents, including commuting to and from work, going out during working hours, going out for lunch during breaks, going to and from class, and picking up and dropping off customers. The design of the survey route follows the following principles: the selected survey time is working days without special major events, the survey duration is the same as the commuting time, and the total survey time exceeds 8 h. The travel plan is based on actual commuting behavior, for example, arrive at the workplace before 1 o'clock, close the app after arriving at the destination, simulate working indoors, and start simulating a home trip at approximately 5 pm. Investigators on the same route try to take vehicles at different times to avoid artificially high data similarity. To ensure the amount of information and processing value of the collected data, each route includes three or more travel endpoints, and as many transportation modes as possible are combined in accordance with the travel mode choices of actual residents. Two modes of transportation (such as subway + bus, etc.) or a different mode of transportation is used for each trip.

Based on the principles, five commuting routes were designed to simulate the actual routes of normal staff, teachers, doctors and other occupations. According to the general commuting rules of Chengdu residents, the residence is set outside the second ring road, the workplace is located in the city, and several lines are set up from different directions in Chengdu, covering the investigation of the traffic status of multiple arterial roads. The design of the simulated commuting route is shown in Table 6.1.

The noncommuting route simulates traffic behavior in the daily life of Chengdu residents on weekends and outings for leisure, including daily life, culture, entertainment and leisure, picking up people, going to a doctor, and business. The design of the survey route follows the following principles: the selected survey time is the weekend; the survey duration is not limited, but to ensure a sufficient amount of data, the survey time should exceed 6 h when possible, and outdoor walking data should be recorded; the purpose of noncommuting trips should also be specified in advance. Based on the above principles, 6 noncommuting routes covering several major business districts in the central city of Chengdu were designed, and the main

Table 6.1 Commuting route design

Travel purpose	Travel route
Home—work—business meeting—meeting guests and returning to work—going home	Donghui Garden 4th District—Tianfu International Financial Center—Global Center—Tianfu International Financial Center—Donghui Garden 4th District
Home—work—go out for a meeting—return to work after the meeting—go home	Jiaogui Lane—Qingyang District Government—Jinjiang Hotel—Qingyang District Government—Jiaogui Alley
Home—work—lunch—return to work after lunch—go home	Vanke·City of Charm—Sichuan Provincial People's Hospital—near Sichuan Broadcasting University—Sichuan Provincial People's Hospital—Vanke·City of Charm
Home—class (morning)—coming home for lunch—class (afternoon)—going home	Sichuan University (West China Campus)—Sichuan University (Wangjiang Campus)—Sichuan University (West China Campus)—Sichuan University (Wangjiang Campus)—Sichuan University (West China Campus)
Home—work—pick up customers—send customers to hotel—go home	Southwest Jiaotong University—Youth Palace Building—North Railway Station—Yuhao Roman Hotel—Southwest Jiaotong University

noncommuting behaviors were simulated, which is of great significance for the analysis of leisure travel behavior. The simulated noncommuting route design is shown in Table 6.2.

Through the design and investigation of these 11 routes, the general travel laws of Chengdu residents can be simulated, the traffic status of Chengdu's main arterial roads at different times can be obtained, and the travel parameters of travelers can be extracted by processing mobile phone sensor data.

6.1.2 Travel Plan for Multiple Modes

Different transportation modes produce different mobile phone sensor data, and efficient individual travel information extraction algorithms should be able to accurately identify the data laws of different transportation modes. To obtain diversified raw data and fully mine mobile phone sensor data characteristics under different travel conditions, not only diversified routes but also the data characteristics corresponding to different methods must be analyzed. In real life, the individual travel of residents typically involves multiple modes of travel. The experimental route should therefore cover the main modes of transportation, including walking, bicycles, cars, buses, and subways. Moreover, the rules of transfer between different modes should be

Table 6.2 Noncommuting route design

Travel purpose	Travel route
Home—grocery shopping—go home—seek medical treatment—go home	Huayu·Jincheng Famous City—Jiaogui Alley—Huayu·Jincheng Famous City—China Railway Second Bureau Group Hospital—Huayu·Jincheng Famous City
Home—go to the park—eat—go home	The famous capital of Huayu·Jincheng—Huanhuaxi Park—dining nearby—the famous capital of Huayu·Jincheng
Home—go to the zoo to play—shopping in the mall—go home	Shuimu Guanghua Community—Chengdu Zoo—SM Plaza (Second Ring Road)—Shuimu Guanghua Community
Home—shopping in the mall—shopping in the mall—going home	North Gate of Southwest Jiaotong University—Chunxi Road—Raffles City Chengdu—Southwest Jiaotong University
Home—shopping in the mall—shopping in the mall—going home	North Railway Station—Chunxi Road—Chengdu China Resources Vientiane City—Southwest Jiaotong University
Home—visiting friends—dining—shopping in the mall—dining—go home	North Gate of Southwest Jiaotong University—Green Shangyuan Lishe on Hengde Road—Xinghui Middle Road—Jinniu Wanda—Zhangjiaxiang—North Gate of Southwest Jiaotong University

considered. This subsection comprehensively considers data collection factors and practical factors, adding 2 survey routes based on the 11 survey routes designed for different travel purposes in 6.1.1 and designing a multimode combined travel plan for these 13 routes.

Based on the subway line and empirical judgment, two additional survey lines were selected. Due to their large travel volume, complex traffic flow composition, and numerous transportation options, these routes have advantages in collecting mobile phone sensor data for multimode travel. The travel routes are as follows.

(1) The Renmin North Road—South Railway Station line running from north to south, starting from the North First Ring Road and Renmin North Road Bus Station in the north to the South Central Ring Road South Station in the south. The total length is approximately 11 km, and this is the main road of the city. This line includes public transportation lines such as Metro Line 1 and Bus No. 16, passing through commercial centers such as Tianfu Square and Jinjiang Hotel. As shown in Fig. 6.1.

(2) Shuhan Road—Shudu Avenue—East Avenue—Dongda Road, going east—west, starting from the Second Ring Shuhan Road Green Garden in the east and going to Dongda Road Huanmao Plaza in the west. The total distance is approximately 10.7 km along Shuhan Road and Shudu Avenue. Both Dongdajie and Dongda Road are important traffic arteries. Moreover, this line includes Metro Line 2, Bus 43 and other bus lines, passing through the People's Park,

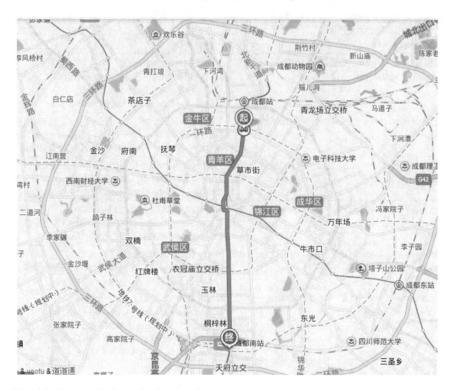

Fig. 6.1 Renmin North Road—South Railway Station Line

Tianfu Square, Chengdu University of Traditional Chinese Medicine Affiliated Hospital and other densely crowded areas. The route is shown in Fig. 6.2.

Corresponding travel mode combinations are designed for these 13 routes. The 5 modes are represented by their English initials: W (walk), A (autobus), C (car), B (bike), M (metro), and the specific combinations are as follows: W-A-W, W-C-W, W-B-W, W-M-W, W-A-A-W, W-A-W-A-W, W-A-M-W, and W-B-W-A-W. The locations of transfer points are set according to the characteristics of each line, such as the People's Park bus station, Jinjiang Hotel bus station, and Rennan Interchange east station, to facilitate subsequent research on travel transfer points, travel OD and other information. This survey plan is rich and covers Chengdu. The data of all types of travel choices of city residents are very representative. Since the survey plans for all lines are essentially the same, this section takes Renmin North Road—South Railway Station as an example. The combined travel survey plan for multiple transportation modes is shown in Table 6.3.

Fig. 6.2 Shuhan Road—Dongda Road Line

6.1.3 Travel Plan for Different Traffic Conditions

The characteristics of the corresponding collected data under different traffic conditions vary substantially. Different input attributes, such as average velocity and velocity variance, affect the algorithm. An algorithm trained in the unblocked state may not be suitable for the congested state. The characteristics of the velocity curve in the smooth state and congested state on the same road section indicates that in the smooth state, the acceleration and deceleration of the vehicle is small, and the vehicle drives stably; in the congestion state, the acceleration and deceleration are considerable. Because it is immobile for a long time, the car takes more time to drive the same road section in heavy traffic conditions.

To verify the applicability of the method and evaluate the effectiveness of the algorithm under different conditions, it is necessary to analyze data under smooth and congested traffic conditions. According to the 13 routes designed in Sects. 6.1.2 and 6.1.3 with different modes of travel, 30 trips are collected for each mode combination on each route: 10 trips are recorded during peak congestion in the morning and evening hours (7:00–9:00, 17:00–19:00) and 20 trips are recorded during normal periods. The processing of congestion data during peak hours is explained in Chap. 7.

Table 6.3 GPS data survey plan for Renmin North Road—South Railway Station

Combination of modes	Route arrangement
Walking-bus-bus-walking (same platform)	Walk from Sun Plaza—Renbei Bus Station (No. 16)—Rennan Sect. 1 (No. 118 in Tongtai)—South Railway Station Bus Station—Walk to Hengdian Film City
Walking-bus-walking-bus- walking (different station)	Walk to Sun Plaza—Renbei Bus Station (No. 16)—Rennan Sect. 1—Walk to Jinjiang Hotel (No. 118)—South Railway Station—Walk to Hengdian Film City
Walking-metro-walking	Walk to Sun Plaza—Renbei Subway Station (Metro)—South Railway Station—Walk to Hengdian Film City
Walking-bus-walking-metro-walking	Walk to Sun Plaza—Renbei Bus Station (No. 16)—Rennan Sect. 1—Walk to Jinjiang Hotel Metro Station—South Railway Station—Walk to Hengdian Film City
Walking-taxi-metro-walking	Walk to Sun Plaza—Renbei Bus Station (Taxi)—Jinjiang Hotel Metro Station—South Railway Station—Walk to Hengdian Film City
Bike-bus-walking	North Railway Station (bike)—Renmin North Road Bus Station (No. 16)—South Railway Station—Walk to Hengdian Film City
Bike-metro-walking	North Railway Station (bike)—Renmin North Road Metro Station—South Railway Station—Walk to Hengdian Film City

6.1.4 Travel Log Collection

To ensure that the data are understood during data processing and that the authenticity of the survey data is checked, it is necessary to assist in recording the travel log during the survey. The travel log records detailed real data, including arrival time, transportation mode, transportation mode change time, location, vehicle parking time, location, and parking reason (such as signalized intersections, bus stops, etc.) that are included in the actual travel process. The recording principle is that the location of the endpoint and the time of arrival and departure must be clearly stated, and the time recording is accurate to the second. Locations of long stays are marked (divided into intersections and congestion), and the cause of the congestion is analyzed. Additionally, for bus routes, bus stops are recorded.

According to the above principles and the collected data format, the travel log format of the survey record is shown in Table 6.4.

Table 6.4 Survey travel log format

Mode	Location and status	Time
Walk	Start of walking	9:01:25
Metro	Enter the subway station	9:17:25
Walk	Exit the subway station	9:42:42
Walk	To Taisheng Road Intersection Station	9:46:40
Bus	Get on the bus (No. 18)	10:09:15
Bus	Arrive at the intersection	10:09:28
Bus	Leave the intersection	10:09:47
Bus	Arrive at Taisheng North Road Station	10:11:12
Bus	Depart from Taisheng North Road Station	10:11:30
Bus	Arrive at Caojiaxiang Station	10:12:48
Bus	Depart from Caojiaxiang Station	10:13:01
Bus	Arrive at the intersection	10:13:32
Bus	Leave the intersection	10:13:49
Bus	Arrive at Ma'an North Road Station	10:15:04
Bus	Depart from Ma'an North Road Station	10:15:24

6.2 Empirical Study of Trip End Recognition Based on Spatial Clustering Algorithm

The identification of travel endpoints based on mobile phone GPS data is the basis for extracting information about transportation modes and travel purposes. Related researchers identify travel endpoints via specific rules or density—based spatial clustering algorithms. In density—based spatial clustering analysis, repeated routes and multiple travel endpoints at the same location are not well recognized; moreover, the interference of being immobile in traffic jams or traffic control also complicates the identification of travel endpoints.

This section analyzes the feasibility of the algorithm based on the method of identifying the travel endpoint proposed in Chap. 3 through examples: the first step is to supplement data preprocessing and partial signal missing trajectories; the second step is to use the ST—DBSCAN. To extract travel endpoint information, the third step is to combine mobile phone three-axis acceleration data to eliminate misrecognition caused by motorized stays, such as traffic jams and traffic control.

6.2.1 Model Parameter Configuration

This section discusses the calibration of the three parameters of travel endpoint recognition by the spatial clustering algorithm. To use a spatiotemporal clustering algorithm to process individual travel GPS trajectory data, it is necessary to define the three parameters of proximity distance $Eps, \Delta T$ neighborhood and corresponding sample points $MinPts$ in the field to determine the spatiotemporal density of core points and corresponding point cluster expansion conditions.

This study analyzed the cumulative probability distribution of the staying time of various types of travel endpoints. When the staying time of the travel endpoint is 553 s, the cumulative probability distribution reaches 95%; that is, 95% of the travel endpoints can be identified by a stay time of 553 s, which meets the basic requirements of the point cluster extension of the spatiotemporal clustering algorithm, as shown in Fig. 6.3. Furthermore, the distance between each track point and its 553rd neighboring track point is shown in Fig. 6.4. The broken line graph has an inflection point (at 162 m): the broken line before the inflection point is relatively flat, the distance is short, the broken line suddenly rises after the inflection point, and the distance is large because the trajectory of the individual travel chain is at the end of the trip. Under these conditions, the distance between track points is more obvious; that is, the individual is always in motion during the trip, and the distance between the track

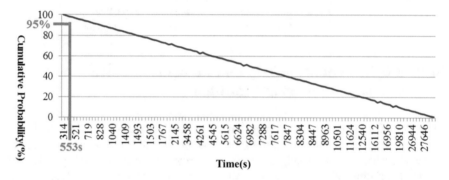

Fig. 6.3 Cumulative probability distribution of stay time at the end of the trip

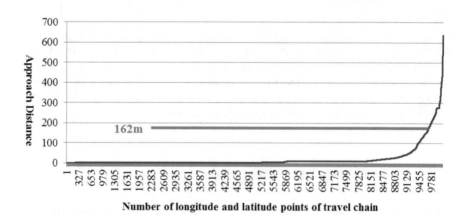

Fig. 6.4 Line graph of the distance from the 553rd point in the sample library

points is larger. At the end of a trip, the individual is immobile, and the distance between track points is smaller. Since this section defines the core points of the spatiotemporal clustering algorithm, considering the time attributes of the points contained in the neighborhood, the time difference should be consistent with the number of points defining the core points. Therefore, the corresponding parameter calibration proposed in this paper is $Eps = 162$ m, $MinPts = 553$, and $\Delta T = 553$.

6.2.2 A Case Study of Trip End Recognition and Travel Trajectory Cutting

The individual travel endpoint identification experiment uses the all—day travel data collected in the previous article. These data include various types of travel in the daily life of Chengdu residents, and the travel modes include walking, bicycles, private cars, taxis, buses, subways and other common transportation modes. The travel endpoint identification method recognizes traffic jams, and travel data under traffic control conditions are collected in a targeted manner. This section uses a travel example to introduce the process of travel endpoint identification and travel trajectory cutting.

Personal cell phone sensor data throughout the day can be collected and exported to the mobile app Xingyi. Figure 6.5 shows the trajectory data of a travel chain collected in Chengdu, as shown by the yellow dots on the map. The yellow dots indicate the GPS track points of the mobile phone collected during travel. Numbers and arrows are used to highlight the travel endpoint and direction of the route. ①, ②, and ③ represent the end points of the trip, and the travel route is preset.

Before the spatiotemporal clustering algorithm is used to identify the travel endpoints, preprocessing of the travel chain is required. Preprocessing is used to address two types issues: ① positioning data with less than 4 visible satellites are removed to avoid errors caused by GPS positioning, reduce problems caused by

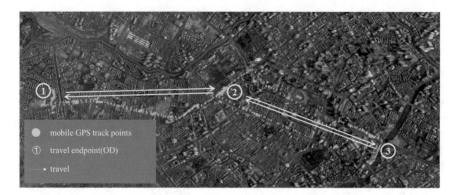

Fig. 6.5 GPS trajectories of mobile phones for individuals traveling throughout the day

mobile phone failures are removed, and instantaneous velocity changes greater than 25 km/h are deleted; ② missing data due to the lack of GPS signal are addressed by calculating the average distance between the track point half a minute before the signal loss and the track point half a minute after the loss. If the average distance is less than 800 m, the track can be regarded as entering and leaving a room. This research supplements this kind of missing signal data by calculating the center position of the trajectory points for 30 s before and after the signal loss, taking the two center positions before and after the signal loss as endpoints, setting the signal loss duration (in seconds) as the number of trajectory points, and filling the corresponding number of track points according to the spacing.

A spatiotemporal clustering algorithm is used to perform spatiotemporal clustering analysis on the preprocessed trajectory data, and clusters of points can be used as potential travel endpoints. Based on the identification of travel endpoint information by the spatiotemporal clustering algorithm, the start and end times of each travel endpoint are obtained, and the time between each pair of adjacent travel endpoints is calculated. If the duration is less than 5 min, the two travel endpoints are merged into a travel endpoint.

According to the travel endpoint identification algorithm proposed in Sect. 3.2 of this book, the mobile phone GPS data of the example travel chain and the travel endpoint are identified for travel endpoints. The identification results are shown in Fig. 6.6a, the blue dots are GPS trajectory points and the red dots are travel endpoints. The blue dots represent trajectory points, and the red dots represent identified travel endpoints. Several travel endpoints are identified with red dots on the map, and the corresponding time periods are extracted, as shown in Fig. 6.6b. The travel endpoint time period recorded in the travel log is shown in Fig. 6.6c, and the accuracy of the algorithm is evaluated by calculating the percentage of correctly identified travel endpoints.

6.2.3   Results and Error Analysis

The spatiotemporal clustering algorithm is used to calculate the recall and error rates after the end points of the row are identified. The recognition results are shown in Table 6.5. Figure 6.7 shows that 5873 of the 6660 real travel endpoints are correctly identified by the spatiotemporal clustering algorithm. The other 787 travel endpoints represent mainly short-term stay behavior, such as picking up people from a certain location, sending children to school, stopping for a short time, and going to work immediately after a short stop. A total of 1320 travel endpoints were misidentified, including points with long-term waiting and serious congestion. One travel endpoint was identified as multiple endpoints. A total of 88.18% of the travel endpoints can be identified by the algorithm; 19.82% of travel endpoint identification errors will not have a significant impact on the extraction of personal travel information based on mobile phone GPS data.

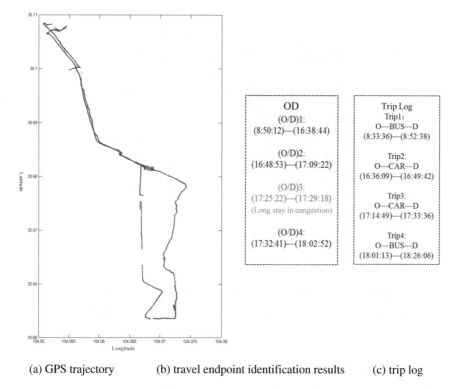

(a) GPS trajectory   (b) travel endpoint identification results   (c) trip log

Fig. 6.6 Travel endpoint identification and GPS track check of the personal travel chain

Table 6.5 Recognition results of travel endpoints

Total number of travel endpoints	Correct recognition number	Number of misidentifications	Correct match rate (%)	False match rate (%)
6660	5873	1320	88.18	19.82

The effect of travel endpoint recognition can be simplified as shown in Fig. 6.7. The left circle is the actual number of travel endpoints recorded in the travel log, and the right circle is the total number of travel endpoints identified by the spatiotemporal clustering algorithm. The intersection of the two circles is the correct identification point.

In the future, a city's resident travel survey data can be used to calibrate the parameters of the spatiotemporal clustering algorithm and analyze the sensitivity of the parameters simultaneously. Under the condition of signal loss, mobile phone signaling data can be collected for supplementary identification; to eliminate the misrecognition of travel endpoints under traffic congestion or traffic control conditions, in addition to the three-axis acceleration data of the mobile phone, the GIS road network database can be used for map matching and recognition.

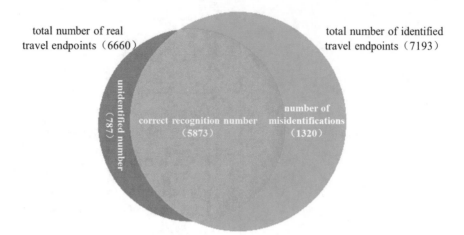

Fig. 6.7 Diagram of travel endpoint recognition results

6.3   Empirical Study of Mode Transfer Point Recognition Based on Wavelet Transform Modulus Maximum Algorithm

With the rapid development of the urban population, economy, travel distance and travel range of travelers, a combination of multiple modes of transportation is often required to complete a trip. Moreover, the connections between different modes of transportation greatly affect the quality of transportation. Reducing transfer time and ensuring convenient and safe transfers are major tasks in transportation planning. The spatial clustering algorithm introduced in Sect. 6.2 can identify individual travel endpoints for multiple trips in a day but fails to identify the transfer point of a single trip. Thus, a method to identify transfer points based on mobile phone GPS data is still needed. This research proposes the application of the wavelet transform modulus maximum algorithm to recognize transportation mode transfer points.

6.3.1   Model Parameter Configuration

The wavelet function, vanishing moment order, and wavelet analysis scale are parameters that need to be configured when the wavelet transform modulus maximum algorithm is used to identify transportation mode transfer points. In this study, the MATLAB wavelet analysis toolbox was used to process travel velocity data.

An appropriate choice of wavelet function is important for the recognition of singular points, and the accuracy of the time recognition of singular points should be taken into consideration. According to the aforementioned principles, the application effects of different wavelet functions are tested, including the Haar wavelet,

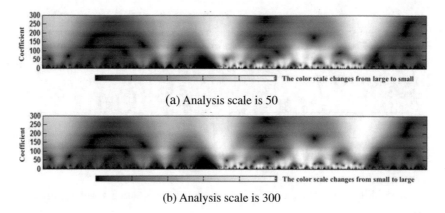

(a) Analysis scale is 50

(b) Analysis scale is 300

Fig. 6.8 Wavelet analysis coefficient modulus calculation results under different analysis scales

Daubechies wavelet and Gaussian function. The experiment finds that the complex Gaussian function performs best in transportation mode transfer point recognition.

The vanishing moment order determines the number of modulus maxima. For a combination of multiple transportation modes for residents, a transportation transfer can usually be understood as an instantaneous completion at a certain location, and the modulus maximum line corresponds to a point in time. Therefore, when choosing the wavelet function, the corresponding vanishing moment of the function should be as small as possible, preferably the first-order vanishing moment function.

The recognition accuracy varies at different scales. An excessively small scale will lead to overrecognition of transfer points; an overly large scale will cause failed recognition. Additionally, there is a need for small-scale accurate identification and large-scale noise reduction optimization. The wavelet analysis coefficient modulus of a sample velocity curve was calculated under different analysis scales, and the results are shown in Fig. 6.8. The analysis scale of Fig. 6.8a ranges from 1 to 50, and the analysis scale of Fig. 6.8b ranges from 1 to 300. The number of modulus maxima needed to obtain the maximum value is different, and the smaller the scale is, the more positions required to obtain the modulus maxima. According to the actual travel situation calibration, when the analysis scale is 300–500, the transfer point identification accuracy is higher.

6.3.2 A Case Study of Mode Transfer Point Recognition

Travel via multiple transportation modes is an important feature of urban residents' travel in China. The accuracy of the identification of transfer points is of great significance to verify the applicability of the recognition algorithm proposed in this study. This section uses collected trip data as an example, draws the modulus maximum line through different analysis scales, identifies the transfer point of the trip, and analyzes

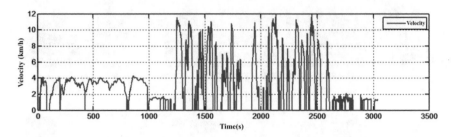

Fig. 6.9 Sample velocity broken line after preprocessing

the applicability of the wavelet transform modulus maximum algorithm introduced above.

A W-B-A-W combination trip is selected as an example, and preprocessing, such as filling missing values and correcting excessive velocity on the velocity polyline, is performed, as shown in Fig. 6.9.

The analysis scale ranged from 1 to 50, and the transfer point of the sample velocity curve was identified. The results are shown in Fig. 6.10. Figure 6.10a shows the original GPS velocity line graph of the W-B-A-W sample data; Fig. 6.10b shows the calculation results of the wavelet transform coefficient modulus for this example under different analysis scales. The modulus of the wavelet transform coefficient at the transfer point of the transportation mode is obviously larger than that of other locations (higher brightness). Figure 6.10c is the wavelet transform coefficient modulus curve obtained when the scale of Fig. 6.10b is 25. The position of the maximum value in the figure is the position of the transfer point under this scale. Figure 6.10d shows the connection of the modulus maximum point corresponding to each scale within the analysis scale (modulus maximum value line). The intersection of this line and the time axis is the identified transfer point. Comparison of the travel logs indicates that in Fig. 6.10d, parking at an intersection or bus stop is misidentified as a transfer in many cases, and further optimization is needed.

The analysis scale is selected from 1 to 300, the sample is reidentified, and the results that appear at the same time as Fig. 6.10 are selected. These modulus maximum value lines are accurate transportation mode transfer marking lines, and the error correction results are shown in Fig. 6.11. The combination of W-B-A-W travel data was divided into 5 travel segments according to different modes of transportation, and each travel segment contained only one mode of transportation. The travel log data shows that at a small scale, the modulus maximum value of the wavelet transform can accurately identify the transfer point of the transportation mode. As the analysis scale increases, the recognition of the transfer point by the modulus maximum value has a time deviation. Taking the intersection of the modulus maximum line and the time axis as the transfer point of the transportation mode, the recognition accuracy can reach the maximum.

According to the same principles, the combination of multiple common transportation modes for residents' travel is recognized, and the recognition result is shown in Fig. 6.12. Figure 6.12a is an example of W-C-W; Fig. 6.12b is a W-M-W

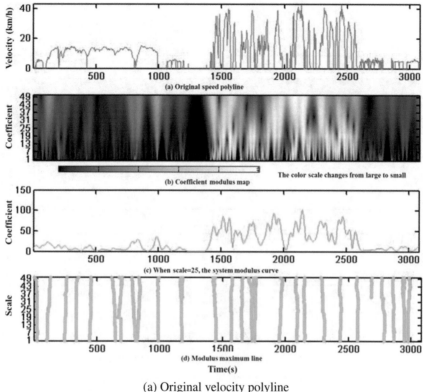

(a) Original velocity polyline

(b) coefficient modulus map at different scales

(c) the system modulus curve when scale=25

(d) recognition result of the modulus maximum line at the transfer point of the mode

Fig. 6.10 The recognition result of the transfer point of the sample transportation mode when the analysis scale is 50

example; Fig. 6.12c is a W-A-W-A-W example; Fig. 6.12d is a W-A-A-W example. The modulus maximum algorithm can identify transfer points of different combinations of transportation modes, and the specific identification accuracy is analyzed in detail in Sect. 6.3.3.

6.3.3 Results and Error Analysis

All the test data were used to identify the time point of the transportation mode transfer, and the error statistics are shown in Table 6.6. Since walking is usually the connection between two modes of transportation, different modes of travel can be

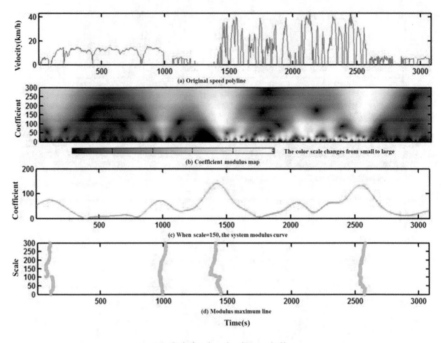

(a) Original velocity polyline

(b) coefficient modulus map at different scales

(c) the system modulus curve when scale=25

(d) recognition result of the modulus maximum line at the transfer point of the mode

Fig. 6.11 Correction results of recognition error of the transfer point of the sample transportation mode

divided into the mode combinations described in Table 6.6. The statistical results indicate that the wavelet transform modulus maximum algorithm can accurately identify the transportation mode transfer time point: the proportion of unrecognized or incorrectly recognized transfer points (error rate) in all combined modes is less than 5%, and the average recognition error at the time of transfer is less than 1 min.

In the traditional questionnaire survey method, the interviewee recalls the time of transfer in the past few days, the time point memory is usually vague, and the magnitude of the error may be as high as half an hour. In addition, according to the statistical results of the error column in Table 6.6, unrecognized transfer points usually occur between walking and bicycle transfers because of the short travel time of the pedestrian section when the residents travel with a combination of walking and bicycle. For example, during a shopping trip, someone walks to the garage first, then rides a bicycle to the parking lot near the mall, and finally walks to the mall for shopping (that is, walk-bicycle-walk). Both walks during the period are used to access the bicycle. If it continues, the time is very short, and obvious singularities

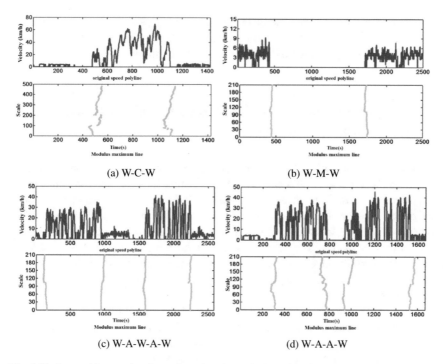

Fig. 6.12 Recognition results of transfer points for typical combined travel modes

Table 6.6 Statistical results of identification errors at transfer time points

Transfer method combination	Absolute error distribution (%)					Average delay (s)
	0–60 s	60–120 s	120–180 s	>180 s	Error rate	
Walking—subway	90.0	7.0	3.0	0.0	0.0	23.3
Subway—walk	90.0	5.0	4.0	1.0	0.0	26.2
Walk—bus	81.2	11.9	5.9	0.0	1.0	31.0
Bus—walk	63.5	10.6	21.2	4.7	0.0	58.5
Walk—bike	92.0	3.3	2.4	0.0	2.3	14.3
Bicycle—walking	77.8	4.0	8.2	5.6	4.4	38.4
Walking—car	55.8	41.2	3.0	0.0	0.0	44.1
Car—walk	74.5	15.6	6.9	2.0	1.0	40.5

may not be generated after wavelet transform. In addition, Table 6.6 shows that the recognition error of the time point of the bus-walking transfer is large, and the error is between 120 and 180 s and the rate is 21.2%. Compared with the travel log data, some of the bus stops of these samples are far from each other. The intersection is very close, and the vehicle continuously stops at the bus station and the intersection for a long time, causing the algorithm to misidentify bus-to-walk transitions, resulting in increased error. However, the bus station is usually close to the stop line at the intersection, which hinders the traffic capacity of the intersection, so the error will be improved with the standard design of the traffic facilities.

6.4 Empirical Study of Travel Mode Recognition Based on Neural Network Algorithm

The traffic mode is an important feature of residents' travel. Identification of the travel mode via mobile phone GPS can provide basic data for studying commuter travel traffic structure, traffic mode selection characteristics of different groups of people, and travel distance characteristics of commuter traffic modes, thereby contributing to urban traffic development planning. Different data characteristics are reflected by different travel modes. The neural network algorithm introduced in Sect. 3.3 can realize data classification. By searching for the relationship between different travel data and output modes, network training can be performed for automatic pattern recognition.

This section empirically analyzes the effect of the neural network algorithm on the input travel section of the traffic mode recognition algorithm and evaluates the feasibility of the proposed algorithm. A total of 70% of the test data are used for model training or calibration, and 30% are used for testing.

6.4.1 Model Parameter Configuration

The travel mode affects the travel trajectory characteristics. For example, the velocity of cars and buses is almost always higher than that of bicycles or walking. On a given route, buses stop more frequently than cars. Therefore, the key link of travel mode identification is to explore the main characteristics that reflect different travel modes. This section configures the relevant parameters of the neural network algorithm to identify traffic modes.

When the neural network algorithm is used to recognize the travel mode, the key parameters that need to be configured are the number of hidden layers, hidden neurons, training algorithm, learning rate, and training period.

A neural network generally includes an input layer, an intermediate layer (also called a hidden layer), and an output layer. A single-layer feedforward neural network can solve only linearly separable problems: a multilayer neural network with hidden layers must be used to solve nonlinear problems. In actual operation, due to the high accuracy of the 3-layer structure and the substantial amount of calculation for the 4-layer structure and above, the 3-layer structure (hidden layer is 1 layer) is selected for training.

Table 6.7 Neural network algorithm configuration

Model	Parameter	Value
Neural networks	Hidden layers	1
	Hidden layer neuron	45
	Training algorithm	Levenberg–Marquardt algorithm
	Learning rate	0.05
	Training time	3000

In the BP neural network, the choice of the number of hidden layer nodes is very important. At present, there is no scientific and universal method for the selection. The most basic principle for determining the number of hidden layer nodes is to use as compact a structure as possible under the premise of satisfying the accuracy requirements, that is, to select as few hidden layer nodes as possible. (1) The number of hidden layer nodes must be less than $N - 1$ (where N is the number of training samples), and the number of nodes (number of variables) in the input layer must be less than $N - 1$. (2) The number of training samples must be more than the connection weight of the network model, generally 2–10 times. According to the above two principles, the number of nodes is approximately 40–50. Since the accuracy is does not differ substantially, 45 is selected as the number of hidden layer nodes for training.

The training algorithm, learning rate and training duration are parameters that determine the complexity of a neural network algorithm. Based on previous research experience and actual operating results, it is found that the calculation time is affected and the accuracy is not significant, so the toolbox default value is used. The specific configuration values are shown in Table 6.7.

In the introduction to the neural network algorithm mentioned above, the four attributes, the average velocity per minute, the maximum velocity, the standard deviation of the travel velocity per minute, and the variance of the acceleration per minute, are used as the input attributes for neural network training. Before the training process of the neural network algorithm, 70% of the collected data are selected as the training data, and the travel mode of each data point is calibrated separately. To analyze in the subsequent model algorithm, the transportation mode is converted into data (Table 6.8): walking is represented by 1, bicycle is represented by 2, bus is represented by 3, and the car is represented by 4. Moreover, the traffic mode is calibrated for each data point, thereby calibrating the key point of the travel trajectory, that is, the traffic mode of the traffic mode conversion point.

Table 6.8 Data conversion of traffic mode

Model	Walk	Bicycle	Bus	Car
Representative value	1	2	3	4

6.4.2 A Case Study of Traffic Mode Recognition

Since the mode of transportation affects the other travel elements and because the data collected by mobile phones cannot directly obtain the mode of travel, mode identification is an important part of travel data processing. This section uses examples to show the neural network algorithm for the identification process of traffic modes by training on some data, identifying other data and outputting the results.

All mobile phone sensor data collected on a route in the design plan were selected, and 70% of the data were taken as training data. Figure 6.13 shows the mean square error between several types of data (training data, test data, validation data) during the training process to evaluate the degree of training.

After 70% of the data are used for training, the trained network is obtained, and the remaining 30% are used as recognition data. The recognition result is determined according to the frequency of the recognition mode. The mode that appears most frequently after recognition within a certain period of time is deemed to be the current mode of transportation. Figures 6.14 and 6.15 show the mode recognition effect of a section of W-A-W and a section of W-B-W travel patterns, respectively. The figure shows that the mobile phone GPS velocity data reflect the representative value (Table 6.8) after the neural network traffic mode is recognized, indicating the corresponding traffic mode in the corresponding time period, and the trend is the same as that of the original velocity curve.

By comparing Figs. 6.14 and 6.15, we find that the recognition results of different transportation modes are different. In the actual driving process of the bus, the behavior of passing through signalized intersections or stopping stations causes the velocity of the vehicle to vary greatly, and there are multiple misidentifications (identified as other non-pedestrian transportation modes), which reduce the accuracy of mode recognition. This phenomenon also occurs in the recognition of the car mode. The velocity of bicycle travel has a small range: only the two modes of walking and bicycle are identified, and the accuracy is high.

6.4.3 Results and Error Analysis

According to the example in Sect. 6.4.2, neural network training is performed on all the collected travel data to obtain the mode recognition results. This study uses the method recognition rate to prove the rationality of the model. The mode recognition rate is defined as the success rate of mode recognition. The most frequent rate among the recognition results is used as the final mode, and the success rate is calculated.

Table 6.9 shows the statistical results of the correct rate of traffic mode recognition. The recognition accuracy rate of different modes of transportation exceeds 70%, especially for walking and cycling, where the accuracy is 94.0 and 93.3%. The difference between acceleration and other parameters is obvious, the data quality of the two methods is usually better, and the volatility is small. In addition, due to the

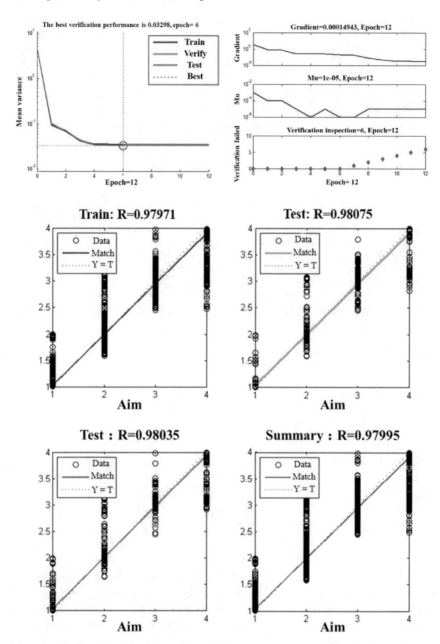

Fig. 6.13 The training process of related parameters

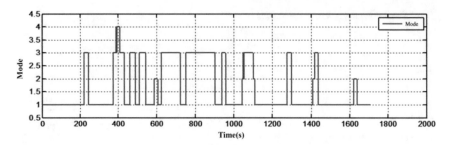

Fig. 6.14 Pedestrian—bus—pedestrian pattern recognition results

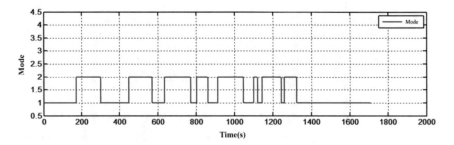

Fig. 6.15 Walking—bicycle—walking pattern recognition results

Table 6.9 Statistical results of the correct rate of traffic mode recognition

	Identified as				Total travel segments	Correctly identified segments	Success rate (%)
	Walk	Bicycle	Bus	Car			
Walk	1034	66	0	0	1100	1034	94.0
Bicycle	7	140	3	0	150	140	93.3
Bus	12	15	260	63	350	260	74.3
Car	1	2	12	35	50	35	70.0
All	1054	223	275	98	1650	1469	89.0

similar travel characteristics of buses and cars, such as travel velocity, the recognition accuracy of the two modes of transportation is relatively similar, 74.3% for buses and 70.0% for cars. As the vehicle stops at a signalized intersection or bus stop, the data characteristics of buses and cars are likely to overlap with the data characteristics of walking or bicycles, resulting in misidentification; that is, buses and cars will be identified as walking or bicycle travel.

The success rate of traffic mode recognition based on a neural network is relatively high. The method can recognize walking, bicycle, and motor vehicles with a high accuracy; the recognition rate of buses and cars is also greatly improved but is still not fully satisfactory. However, the neural network algorithm also has limitations, such as being unable to explain its own reasoning process and reasoning basis, being unable to ask the user necessary inquiries (when the data are insufficient, the neural network cannot work), turning the characteristics of all problems into numbers, and changing all reasoning into numerical calculations may result in a loss of information in the conversion process.

6.5   Empirical Study of Travel Chain Recognition Optimization Based on GIS Map Matching

In the previous data analysis, because the travel characteristics of buses and cars are similar. For example, the maximum velocity, average velocity, travel line and other travel characteristics and velocity all obviously vary, and the mode of transportation may be easily identified as an incorrect mode. Bus and car travel are difficult to distinguish based on only velocity characteristics, so new factors must be considered. In actual travel, the two modes stop at signalized intersections, but buses have regular stopping behavior that differs from that of cars.

According to the operation characteristics of the two modes, this part analyzes the method of bus and car identification based on GIS map matching. The basic idea is to identify key points with large velocity changes, including vehicle stops, and match the data with the longitude and latitude of bus stops on the map according to the matching degree to distinguish between the bus and car modes.

6.5.1   Model Parameter Configuration

According to the map matching algorithm introduced in Sect. 3.5 of Chap. 3, the parameters to be configured are the range domain "R" and ratio threshold "P". Moreover, to match the station location, the longitude and latitude of all bus stops in the survey data must be collected.

The accuracy of bus or car pattern detection depends on the critical point of parameter settings of the matching algorithm. In the process of site matching, the parameters selected for site matching on the map must be configured. According to the matching algorithm mentioned above, we select the range domain "R" and ratio threshold "P" and compare the detection rate for different parameter values for the parameter configuration of "R" and "P".

The experimental results show that as the range domain "R" increases, the detection rate of the bus mode increases, while the detection rate of the car mode decreases. As the ratio threshold "P" increases, the detection rate of the car mode increases, and the detection rate of the bus mode decreases. This finding is consistent with reality: bus stop locations are located in a relatively small range, and the bus stops at each platform on the route. While the parking behavior of a car is relatively random, there is a certain probability of stopping within a certain range of the bus stop; therefore, it can pass the inspection of bus or car stops on the route. The larger the radius of detection is, the fewer the car matching stations and the lower the detection rate. In contrast, when the radius of detection reaches a certain range, the matching stations will be consistent with the actual stops, and the detection rate will be improved.

Tables 6.10 and 6.11 show the bus and car site matching results for different "R" and "P" values. When the matching radius "R" is 60 m and "P" is 70%, the detection accuracy of buses and cars reaches a high level.

The longitude and latitude of the bus stops on the survey route are measured, and the matching degree is determined by the degree of matching between the longitude and latitude of the position and the bus stop. Taking the Renmin North Road South Railway Station as an example, this section lists the longitude and latitude of bus stops in Table 6.12.

Table 6.10 Detection rate of cars under different R and P values

Detection rate (car)	Ratio threshold "P"			
	60%	70%	80%	90%
Radius R (m)				
20	100%	100%	100%	100%
30	100%	100%	100%	100%
40	50%	100%	100%	100%
50	25%	100%	100%	100%
60	25%	100%	100%	100%
70	13%	100%	100%	100%

Table 6.11 Detection rate of buses under different R and P values

Detection rate (public) transport)	Ratio threshold "P"			
	60%	70%	80%	90%
Radius R (m)				
20	36%	0	0	0
30	82%	55%	0	0
40	91%	73%	45%	0
50	91%	82%	73%	0
60	91%	91%	82%	0
70	91%	91%	82%	9%

Table 6.12 Longitude and latitude of bus stops from Renmin North Road to South Railway Station

Site name	Longitude	Latitude
Baliqiao Road	104.0740008	30.70727535
Tofu weir	104.0772494	30.70207560
Beizhan East 2nd Road	104.0771634	30.69788342
North Railway Station East	104.0752127	30.69705072
North of Sect. 2 of Renmin North Road	104.0705950	30.69426500
Renmin North Road	104.0708800	30.68931333
Section 1 of Renmin North Road	104.0698280	30.68433915
South section of Renmin North Road	104.0696000	30.68165167
Wanfu Bridge	104.0683250	30.67876667
Section 2 of Renmin Middle Road	104.0638101	30.67247619
Section 1 of Renmin Middle Road	104.0627782	30.66419211
Section 1 of Renmin South Road	104.0631120	30.65753477
Jinjiang Hotel	104.0635050	30.65100000
Huaxiba	104.0636076	30.64427167
The third section of Renmin South Road	104.0638346	30.64069690
Sichuan Gymnasium	104.0639014	30.63478169
North Sect. 4 of Renmin South Road	104.0640924	30.63039786
Ni Jiaqiao	104.0641312	30.62612719
Tongzilin	104.0643093	30.61809388
South Railway Station East Road West	104.0667876	30.61092100
South Railway Station East Road	104.0725133	30.61081856
Tianren Road South China intersection	104.0735788	30.60404513
Tianren Road	104.0705423	30.60445243
Heping community	104.0679144	30.60470652
Comprehensive transportation hub of South Railway Station	104.0611508	30.60672394

6.5.2 A Case Study of Travel Mode Recognition Optimization

The key step of the method in this paper is the feasibility of key point recognition. If the key point of velocity change can be recognized well, the bus stop can be included for further recognition. In this section, the feasibility of bus and car mode detection based on the site map matching algorithm is verified through case analysis. In this case, the "neural network + map matching" method is selected to identify the traffic mode.

According to the neural network algorithm introduced in 6.4, after mode recognition is completed, the key points generated by the large velocity change will also be identified.

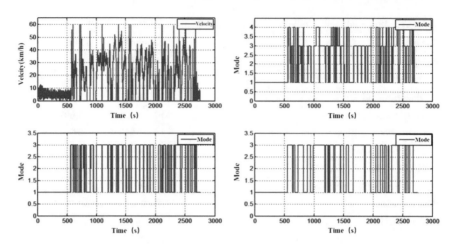

Fig. 6.16 Test results for the walking-bus-walking trip chain

Figure 6.16 shows the key point detection process and results (using a neural network algorithm) for a walking—bus—walking segment. Figure 6.16a shows the original GPS acquisition velocity of the trip; Fig. 6.16b shows the recognition result using a neural network. It is difficult to distinguish the bus and car modes, and the part identified based on the bus (No. 3) velocity is mistakenly identified as car data (No. 4). Figure 6.16c shows the detection results of motor vehicles, including bus and car modes; Fig. 6.16d shows the modified results after correcting some unreasonable mode conversion. As shown in Fig. 6.16d, the proposed method has detected all the key points in the travel example.

Figure 6.17 shows the key point matching and trip information recognition results. Figure 6.17a shows the comparison between the key points and the detected bus stops, in which the blue circle is a key point and the red box is a key point for successful matching. As shown in Fig. 6.17b, the matching result is compared with the actual bus stop locations: the red box is the detected bus stop and the green box is the actual bus stop. The matching rate is greater than 85%. Figure 6.17b, c show the detection results of driving information, that is, walking segment (from 0 to 557 s)—bus segment (from 558 to 2694 s)—walking segment (from 2695 to 2761 s). According to the time of mode conversion, the location (latitude and longitude) of mode conversion can also be extracted from the original data. As shown in Fig. 6.17d, according to the very low matching rate (only one bus stop matches), the proposed method can successfully detect the bus and car modes.

As seen from Figs. 6.18 and 6.19, for the two modes of bus and car, the difference in station matching results is clear: on a given line, the bus stops basically match, while the number of cars matching with bus stops is far less. Thus, bus stops can be used as an important factor to distinguish the two modes.

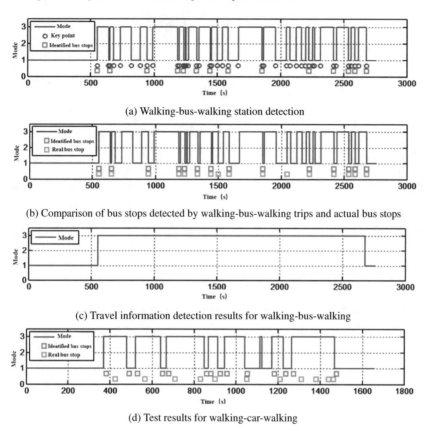

(a) Walking-bus-walking station detection

(b) Comparison of bus stops detected by walking-bus-walking trips and actual bus stops

(c) Travel information detection results for walking-bus-walking

(d) Test results for walking-car-walking

Fig. 6.17 Station matching and traffic information detection results

6.5.3 Results and Error Analysis

On the basis of the results of the original neural network algorithm for traffic mode recognition, combined with the map matching algorithm, the comparison of mode recognition accuracy is shown in Table 6.13. Due to the obvious velocity characteristics of the original walking, bicycle and motor vehicle (including bus and car), the recognition accuracy of the neural network algorithm is high. Furthermore, the map matching algorithm aims to distinguish only bus and car modes, so it has no impact on the recognition accuracy of other travel modes. For the comparison of bus and car modes, the map matching algorithm has no impact on the recognition accuracy of other travel modes. After using the map matching algorithm, the recognition accuracy has changed greatly: compared with the original 74 and 70%, the recognition accuracy of bus and car modes has reached 93.

In summary, the simple machine learning algorithm is not sufficiently accurate to distinguish the bus and car modes, so a map matching method is proposed to improve

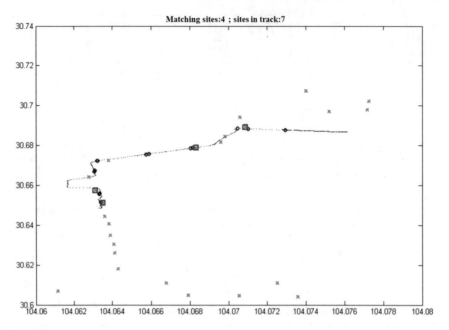

Fig. 6.18 Map matching of car trip stations

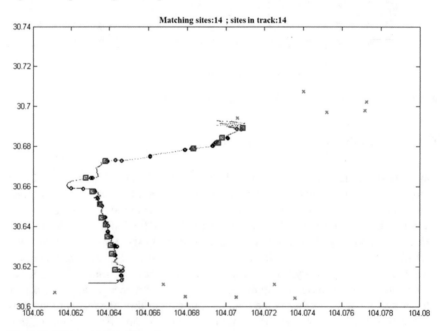

Fig. 6.19 Map matching of bus stops

Travel mode	Mode recognition accuracy (%)	
	Neural network only	Neural network + map matching
Walking	94	94
Bicycle	93	93
Vehicle	95	95
Bus	74	93
Car	70	93
Average	86.2	93.6

Table 6.13 Comparison of statistical results and detection rate using the map matching algorithm

the bus and car recognition accuracy. On the basis of the characteristics of bus stops, the key points in the travel chain are matched with bus stops, and the matching rate is used to distinguish the two modes, greatly improving the identification of buses and cars.

6.6 Summary

In this chapter, through an empirical analysis of Chengdu City based on the mobile phone sensor data mining algorithm proposed in Chap. 3, the difficulties and key technologies of individual travel parameter extraction using mobile sensor data are summarized, the development process of the example is introduced in detail, and the application of a spatial clustering algorithm, wavelet transform modulus maximum algorithm and neural network algorithm is evaluated for processing mobile sensor data. The effect of individual travel parameter extraction in the model is also assessed. The results include the following four aspects:

(1) Using the spatial clustering algorithm and real collected data, the travel endpoint stay time (553 s) and the track point spacing (162 m) can be calibrated to effectively identify travel endpoints from the mobile GPS data. The correct matching rate of travel endpoints reaches 88.18%. Because it is difficult to accurately identify short stops, 19.82% of the endpoints are mistakenly identified as travel endpoints.

(2) The wavelet transform modulus maximum algorithm can be used to identify the transfer points among transportation modes and has a good recognition effect for multiple modes of transportation. The proportion of unrecognized or misidentified transfer points (error column) in all combination modes is less than 5%, and the average recognition error of transfer time points is less than 1 min.

(3) The neural network algorithm can be used for mobile GPS data travel mode recognition, especially for walking and bicycle recognition, and the accuracy

rate is as high as 94.0 and 93.3%. However, for bus and car travel, the information recognition effect is not ideal, and the accuracy rate is 74.3 and 70.0%. The proposed map matching algorithm based on bus stop GIS can compensate for the shortcomings of a single data mining algorithm, and the final bus and car recognition accuracy can be improved to more than 90%.

Chapter 7
Influence Parameters and Sensitivity Analysis

Different from theoretical research, the promotion and application of any new technology is usually affected by the complex environment and technical conditions in the actual environment. The selection of a technical algorithm model, the setting of the data sampling frequency and the actual traffic environment have different degrees of impact on the application of mobile sensor traffic survey technology. The analysis and evaluation of the influence mechanism and effect of these key parameters is an important component of the mature application of the technology. This chapter systematically and comprehensively compares and evaluates the application of mobile phone sensor traffic survey technology with different data mining algorithms, sampling frequencies and traffic states through targeted design of field data collection experiments to provide a reference for the field promotion of the technology. The application provides a technical reference. First, the most popular and efficient data mining algorithms, namely, neural networks, support vector machines, Bayesian networks, random forests, and GIS-based geographic information matching algorithms, are used to evaluate the advantages and disadvantages of the algorithm model for different traffic parameter identification and selection effects. Second, a large number of field data collection experiments were conducted by setting different sampling frequencies (1, 2, 3, 4, 5, 10, 20, 30, 60, 120 s) under different traffic conditions (smooth state, congested state) to analyze the influence of the data sampling frequency and traffic conditions on the application of the technology in practice.

Section 7.1 studies and analyzes the influence mechanism of the algorithm model, data sampling frequency and traffic state on the accuracy of the application. Section 7.2 designs diversified field data collection experiments, introduces the characteristics and fluctuation level of field data collected under different test conditions, and intuitively shows the difference in data under different conditions. Sections 7.3 and 7.4 respectively, compare and evaluate the algorithm model and data sampling frequency. Section 7.5 summarizes the research results of this chapter, refines the research results, and presents guiding suggestions for the practical application of this technology.

© Tongji University Press 2022
F. Yang and Z. Yao, *Travel Behavior Characteristics Analysis Technology Based on Mobile Phone Location Data*, https://doi.org/10.1007/978-981-16-8008-3_7

7.1 Influencing Factors and Mechanism

The model algorithm, data sampling frequency and traffic state are important factors to consider in the practical application of GPS traffic survey technology and play a key role in the effectiveness of the technology.

1. Influence of the Model Algorithm on Accuracy

The computer theory and process of different algorithm models are different. For example, a neural network learns the mathematical relationship between known input attributes and output results through feedback so that it can train and obtain a function of the relationship between different input and output data to identify new data. This approach has a good application effect for multipattern classification problems, but the model training usually requires improvement. A large amount of historical prior data cannot explain the reasoning process and reasoning basis. The support vector machine algorithm can obtain high-dimensional hyperplanes by analyzing the relationship between known input and output data and can then be used for data classification. This approach follows the basic theory of nonlinear mapping and can achieve efficient model "transductive reasoning" with small samples. However, the support vector machine algorithm has some difficulties in multiclassification problems, and the computational time cost is high. Therefore, when different algorithms are applied to traffic trip information recognition, significant differences in technical application effect will occur, especially when the data sampling frequency and traffic state change. Even the application effect of the same algorithm will change significantly, so a model with a better application effect in the smooth traffic state may not be applicable to the data in the congested state. Moreover, the model with significant effects under dense data sampling may not be able to recognize sparse data. The evaluation of the advantages and disadvantages of different algorithms and the applicability of different parameter identification approaches will play a vital role in the practical application of mobile sensor survey technology.

2. The Influence of the Data Sampling Frequency on Accuracy

The data sampling frequency refers to the frequency of data acquisition by mobile phone sensors. At present, the positioning interval of GPS sensors can reach 1 s, and the location frequency of triaxial acceleration sensors can reach 0.2 s. A higher sampling frequency indicates that more data are collected during the trip, the advantage of which is that the data can better reflect the individual travel trajectory and travel characteristics. For example, under low—interval sampling of 1 s, the obtained point velocity data can reflect the state of motion or rest of an individual and the acceleration and deceleration characteristics. However, a high sampling frequency entails a large amount of data acquisition, storage and data calculation, which increases the technical cost considerably. Additionally, the location frequency affects the battery life of mobile phones and has a great influence on the willingness of individuals to participate in the investigation. Therefore, in practical applications, the selection of the data sampling frequency usually must weigh the comprehensive benefits of data precision and investigation cost according to the purpose of the investigation (macro or micro).

The impact of the data sampling frequency on traffic parameter identification is reflected in the large amount of traffic mode identification and transfer time point identification. For traffic mode recognition, theoretically, a low sampling frequency will reduce the accuracy of traffic mode recognition. For example, for a 30-min walking (5 min)—bus (20 min)—walking (5 min) trip, when the sampling interval is 1 s, the three travel segments have 300 points, 1200 points, and 300 points, respectively, and the three segments of data can accurately reflect the combined travel characteristics of the three modes. However, when the sampling interval is reduced to 120 s, the trip data volume becomes 4 points, 10 points, and 4 points, which may be insufficient. In addition, when the amount of data is small, it is particularly difficult to distinguish between buses and cars because some short time or short distance travel information will be omitted.

The sampling frequency also has a significant effect on the transfer time of the traffic mode. Individual daily travel trajectories usually include many stop times, such as traffic mode change points, red light waiting points at intersections, and bus stops. Therefore, the data characteristics of the travel trajectory before and after the stop must be analyzed in detail to identify traffic mode conversion points. In practice, the transportation mode transfer is usually completed in a relatively short time. When the sampling interval is small (such as 5 min), due to the large sampling interval, data points from the transfer period are often not collected, resulting in missing transfer information. In addition, even if the data points in the transfer period are collected, it is difficult to effectively distinguish transfer stops from other stops, such as intersections or platform stops, due to the small amount of travel data at the front and back ends (for example, there are only a few data points in the previous case), and the transfer time point error will thus increase.

3. Influence of Traffic State on Accuracy

The traffic state directly affects the travel data characteristics of different traffic modes. Generally, the running velocity of different traffic modes is a key indicator for traffic mode identification. The walking velocity ranges from 0 to 8 km/h, and the stable velocity is approximately 3–6 km/h; the bicycle velocity ranges from 0 to 17 km/h, and the stable velocity is approximately 10–15 km/h; the bus velocity ranges from 0 to 40 km/h, and the stable velocity is approximately 20–30 km/h; the car velocity range is 0–120 km/h, and it fluctuates greatly according to the different road and traffic conditions. Because the subway is running underground where the mobile GPS signal is shielded, the data cannot be obtained, and its velocity is usually recorded as 0, so subway travel must be identified based on geographical information matching of the upper and lower stations.

Under normal circumstances, identification errors arise when velocity is used for mode identification. When the traffic state changes, the limited operation of each travel mode makes the proportion of the normal operation velocity decrease, and the proportion of the low-velocity overlap area increases, which makes mode identification difficult. Due to traffic congestion, motorized travel modes such as buses and cars will stop and go frequently, and their velocity will slow greatly at various points. At such times, the velocity characteristics of motor vehicles, bicycles and even walking become similar. When using a machine learning algorithm for pattern recognition, the error of pattern recognition will increase. In addition, in the practical application of the technology, whether the algorithm model trained under the unblocked state is also suitable for the congested state and how its accuracy fluctuates must be verified and evaluated in detail to provide a reference for the technical application.

7.2 Data Characteristics Under Different Experiment Conditions

To evaluate the application effect of traffic investigation technology based on mobile phone sensor data with different algorithms, the data sampling frequency and the traffic status, this section designs a field data acquisition test to collect and analyze a large amount of field data, explores the data characteristics and change rules under different conditions, and provides support for the identification and extraction of the following traffic travel chain parameters.

7.2.1 Data Collection

The data acquisition app based on the smartphone sensors developed in Chap. 4 is used for field data acquisition, and the test scope covers 10 urban expressways and trunk roads in downtown Chengdu. Chengdu, the financial, technological and transportation hub in Southwest China and the capital city of Sichuan Province, has a population of more than 14 million, and the number of cars is among the top five in China. Chengdu is a typical large city with a high population density and a complex traffic environment. Therefore, Chengdu represents a good test scenario to verify the application effect of mobile phone sensor traffic investigation technology.

In the daily travel of residents, walking is usually regarded as an intermediate transition mode between two modes of transportation. Any complex multimode traffic combination can be divided into the basic unit of walking-X mode-walking. Therefore, to control the complexity of the experiment, three basic modes are considered: walking-bicycle-walking, walking-bus-walking, and walking-car-walking. The data sampling frequency is set to 1, 2, 3, 4, 5, 10, 20, 30, 60, and 120 s, and the traffic state includes two categories: nonpeak unblocked state and morning and evening peak congestion state. Each mode combination is tested on each test route 30 times, and the morning and evening peak congestion period data account for approximately 1/3 of the total data. Finally, the test data collected in this part constituted 2160 h, close to 7,776,000 records. Approximately 110 undergraduates and graduate students from

Southwest Jiaotong University participated in the field data collection. The experimenters were required to record travel log data of the whole process, including travel start and end time, travel start and end point, all traffic modes used, traffic mode transfer point times, stop times at intersections and bus stops, congested road sections, etc. The travel log data are used to verify the accuracy of the results.

7.2.2 Data Analysis

1. Characteristics of Mobile Phone Sensor Data Under Different Sampling Frequencies

The key to traffic mode identification is to make full use of and mine the differences in travel data of different modes. For example, the maximum walking velocity on urban roads is approximately 10 km/h, while a car can reach 40 km/h. However, different data sampling interval result in significantly different original data densities. For example, for a half-hour multimode trip, a location sampling interval of 1 s generates 1800 data points, while a location sampling frequency of 5 min generates only 6 data points. Clearly, if the location sampling frequency and the amount of data collected are insufficient, error will result in the identification of traffic modes and transfer points. Figure 4.19 in Chap. 4 shows the velocity fluctuation characteristics of cars at different sampling frequencies on the same road section. As the sampling time interval increases, the time interval between two data points increases. When the positioning sampling interval is less than 30 s, the velocity characteristics and fluctuation trends are similar. When the interval is increased to 60 s, the amount of data collected is greatly reduced, and the fluctuation trend is rapidly weakened. Furthermore, when the sampling interval reaches 120 s, it is very difficult to distinguish different modes of transportation due to the lack of data.

2. Data Characteristics of Mobile Phone Sensors in Different Traffic Conditions

Under different levels of traffic congestion, the driving conditions of vehicles will be restricted by the road and traffic flow itself, resulting in different degrees of operation interference. Due to the slow velocity of motor vehicles, their characteristics may appear to be similar to those of bicycles and pedestrians, so it will be difficult to identify the mode of travel. Moreover, it will become more difficult to distinguish between different modes, and for the same mode, the algorithm model trained with unblocked travel data will no longer be suitable for the identification of congestion data, and the application effect will become unsatisfactory.

Figure 4.36 in Chap. 4 shows the fluctuation characteristics of car acceleration data under different road congestion levels, including smooth state, general congestion and severe congestion. As the traffic congestion level increases, the amplitude of the acceleration fluctuation is significantly weakened. The regular fluctuation characteristics in the smooth state cannot be recognized in the serious congestion state.

7.3   Sensitivity Analysis of Travel Mode Recognition

In this part, by comparing the recognition results of the algorithm with the travel log information recorded by the investigators, we use the accuracy index of traffic mode recognition to evaluate the effect of traffic mode recognition and analyze the error.

7.3.1   Influence of Algorithms

Table 7.1 shows the statistical results of traffic mode identification under different algorithms, including a neural network, support vector machine, Bayesian network, and random forest, and the four algorithms combined with GIS map matching comprehensive model. The four machine learning algorithms can be used successfully for traffic mode recognition based on GPS surveys. When only a single machine learning algorithm is used for traffic mode recognition, the recognition accuracy of walking, bicycles and motor vehicles exceeds 90%, and the recognition effect is good. However, the recognition effect of buses and cars is not ideal (60–75%). The main reason is that the travel characteristics data of walking, bicycle and motor vehicle modes are quite different, and the machine learning algorithm can distinguish these three modes, but the characteristics of buses and cars are too similar to distinguish. The application results of the four methods are generally similar, and the support vector machine performs best. Moreover, the proposed map matching approach improves bus and car identification: the recognition rates of the two methods exceed 90%, an increase of 20%-30%.

7.3.2   Influence of Data Sampling Frequency

According to previous research, this section selects the best traffic mode recognition method, i.e., support vector machine + map matching algorithm, to analyze the impact of the sampling frequency. The results of traffic mode identification under different data sampling rates are shown in Fig. 7.1 As the sampling interval increases, the recognition rates of walking, bicycles, buses and cars are gradually reduced. When the sampling frequency is large (1–5 s), the recognition accuracy of the four traffic modes is greater than 80%, especially for the high sampling frequency of 1–3 s, and the recognition accuracy of the four traffic modes exceeds 90%. When the sampling rate is more than 20 s (more than 70%), the accuracy of traffic mode identification decreases rapidly. When the sampling rate is more than 120 s, with the exception of walking (60%), the accuracy of the different modes is approximately 30%. This is mainly because when the positioning frequency is low, the distance between data points on the travel path is long, and many travel characteristics, such as a bus stopping at the platform, cannot be recorded, and the application effect of

Table 7.1 Statistical results of traffic mode recognition accuracy under different algorithms

Trip mode	Pattern recognition accuracy (%)				Machine learning algorithm + graph matching			
	Machine learning algorithm							
	Neural network	Support vector machine	Random forest	Bayesian network	Neural network	Support vector machine	Random forest	Bayesian network
Walking	94	97	96	94	94	97	96	94
Bicycle	93	95	91	94	93	95	93	94
Vehicle	95	96	93	93	95	96	93	93
Bus	74	74	69	72	93	95	90	96
Car	70	73	65	67	93	93	94	93
Average	86.2	87.8	83	84.8	93.6	95.2	93	94

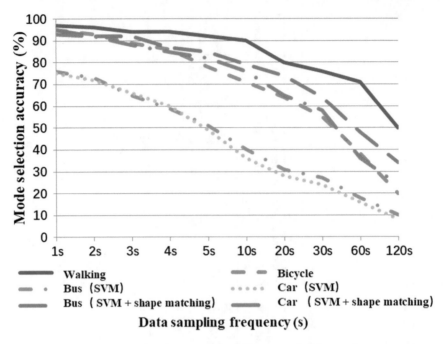

Fig. 7.1 Traffic mode recognition accuracy under different sampling frequencies

map matching becomes worse. With a decrease in the sampling rate, the accuracy of traffic mode recognition is always 20–30% lower than that of the SVM + map matching algorithm. Therefore, in the practical application of GPS survey technology, considering the battery service time and data calculation cost, a sampling interval of 5 s is recommended, and the maximum sampling interval should not be greater than 20 s.

7.3.3 Influence of Traffic Condition

There is a large difference in the operation characteristics between the nonpeak traffic flow and the peak traffic flow. In this part, the support vector machine and map matching algorithm are used to analyze the application effect of peak and the nonpeak travel data. The statistical results under different traffic conditions are shown in Fig. 7.2. The correct rate of pattern recognition for walking and cycling modes is high, reaching more than 90%, regardless of whether traffic is smooth or congested. This is mainly because traffic congestion has little influence on walking and cycling. However, for buses and cars, the recognition when using only the SVM algorithm is approximately 75% when traffic is smooth, while the recognition accuracy drops to 54 and 62% when the traffic is congested due to vehicle deceleration and stop

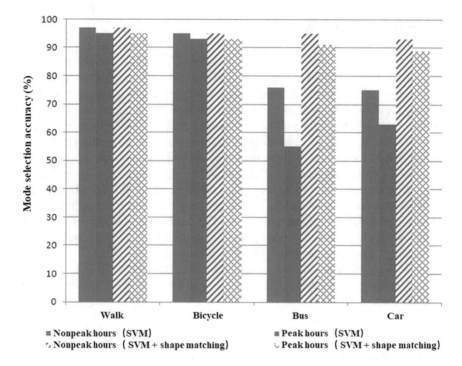

Fig. 7.2 Traffic mode recognition accuracy under different traffic conditions

and go phenomena. Combined with the SVM and map matching algorithm, traffic congestion has little effect on bus and car recognition, and the accuracy of traffic mode recognition is approximately 90% in both states. This is mainly because traffic congestion usually occurs only for part of the entire travel process; although the success rate of matching between car stops and bus stops increases, the bus proportion standard, which is wrongly identified by the map matching algorithm, is usually not met.

7.4 Sensitivity Analysis of Mode Transfer Time Recognition

In this part, the algorithm recognition results are compared with the travel log information, and the evaluation and error analysis are conducted using the identification error index of the transfer time point of the traffic mode.

7.4.1 Influence of Algorithm

The statistical results of the transfer time point identification error of the traffic mode under different algorithms are shown in Table 7.2. The considered algorithms are a comprehensive neural network, support vector machine, Bayesian network, random forest and GIS map matching algorithm, and the data sampling interval is 1 s. The recognition accuracy of the different algorithms for the transfer point of the transportation mode is generally good: the proportion of recognition errors that are less than 20 for walking-bicycle, walking—bus and walking—car is greater than 80%, and the evaluation error is less than 30 s. Compared with that of the traditional questionnaire survey, the accuracy is greatly improved. The application effects of the four algorithms are similar, and the advantages and disadvantages of the algorithms are not obvious. In addition, the figure shows that the transfer point recognition error of the walking-bicycle traffic mode is usually larger than that of the walking-bus and walking-car modes: the proportion of errors that are greater than 30 s is clearly higher. This is mainly because the velocity difference between walking and bicycles is less than that between walking and car/bus, and the difficulty of the algorithm is relatively high.

7.4.2 Influence of Data Sampling Frequency

Figure 7.3 shows the cumulative percentage of the absolute error distribution of transfer points under different sampling frequencies. The support vector machine + map matching is considered. Clearly, the distribution characteristics of the error percentage change greatly under different sampling frequencies. When the sampling interval is high (1–5 s), the proportion of transfer point errors that are within 10 s reaches 50–70%, and the proportion of transfer point errors that are greater than 30 s is within 20%. As the sampling rate decreases, the transfer point error increases significantly. When the positioning sampling interval is greater than 60 s, the proportion of errors that are less than 10 s decreases to less than 20%, while the proportion of errors that are greater than 30 s increases to more than 50%. The overall error trend shows that with decreasing sampling frequency, the data become increasingly sparse, and the identification error increases.

Figure 7.4a–c show boxplots of the error distribution of transfer point recognition under different sampling frequencies. The figure shows five main characteristic elements of the transfer point recognition error: maximum, upper quartile, median, lower quartile and minimum. In addition, the red line represents the mean error under different sampling rates. Figure 7.4 shows that with increasing sampling interval, the maximum value, median value and average value of the identification error increase gradually. When the interval is small, such as 1–5 s, the transfer point error of walking-bicycle is larger than that of walking-bus and walking—car. When

Table 7.2 Statistical results of transfer time point identification error for different algorithms

Method	Trip mode	Absolute error distribution (%)					Absolute error (s)		
		0–10 s	10–20 s	20–30 s	>30 s		Minimum value	Maximum value	Average value
NN + map matching	Walking-bicycle	56.8	18.6	11.5	13.1		1	188	20.6
	Walking-bus	63.5	21.2	7.9	7.4		0	156	16.3
	Walking-car	65.2	23.1	6	5.7		0	148	13.6
SVM + map matching	Walking-bicycle	60.1	20.2	8.6	11.1		0	185	19.7
	Walking-bus	65.5	20.6	4.7	9.2		0	174	16.5
	Walking-car	72.8	11.8	6.5	8.9		0	130	14.3
Random forest + map matching	Walking-bicycle	55.6	19.8	11.4	13.2		2	207	22.4
	Walking-bus	69.1	14.7	6.5	9.7		0	168	17.3
	Walking-car	66.4	20.1	6.1	7.4		0	159	14.7
BN + map matching	Walking-bicycle	57.7	16.5	10.5	15.3		1	221	23.5
	Walking-bus	63.5	20.6	6.3	9.6		0	176	17.1
	Walking-car	67.2	15.5	9	8.3		0	185	15.8

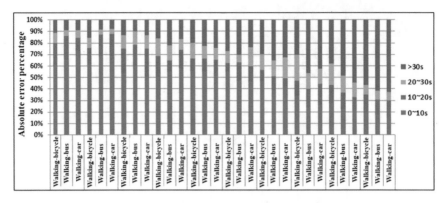

Fig. 7.3 Statistical results of the cumulative percentage of transfer point identification error under different sampling rates

the sampling frequency is greater than 5 s, the transfer point errors of walking-bus and walking-car are higher than those of walking—bicycle. The sensitivity of the sampling frequency to walking—bus and walking—car is higher than that of walking-bicycle. Generally, when the positioning interval is not greater than 10 s, the average error of the three types of transfer points is less than one minute; when the positioning frequency increases to 120 s, the average error is approximately three minutes.

7.4.3 Influence of Traffic Condition

In this section, the support vector machine and map matching algorithm are used to analyze the transfer point identification error of the travel data in the peak period and nonpeak period. The identification results of traffic mode transfer points during peak and nonpeak travel periods are shown in Table 7.3. The transfer point identification error of the traffic mode is much higher under traffic congestion, especially the transfer point identification error of walking-bus and walking-car. The average walking-bus error increases from 16.5 to 57.4%, and the average walking-car error increases from 14.3 to 67.8%. In terms of the absolute error distribution, the proportion of errors within 0–10 s is reduced by 20–30%, and the proportion of errors greater than 30 s is increased by nearly 20%. The impact of traffic congestion on the walking—bicycle buffer point recognition error is small. As mentioned earlier, this is mainly because traffic congestion has less impact on walking and cycling than on bus and car travel.

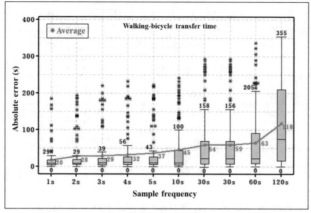

(a) Walking-bicycle

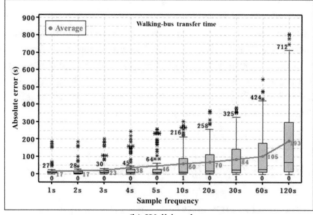

(b) Walking-bus

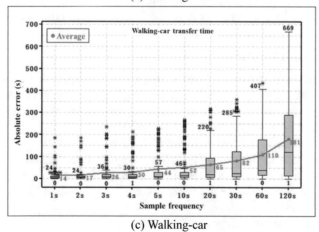

(c) Walking-car

Fig. 7.4 Boxplot of transfer point identification error under different sampling frequencies

Table 7.3 Identification results of transfer points of travel modes in normal and peak travel periods

Traffic conditions	Trip mode	Absolute error distribution (%)				Absolute error (s)		
		0–10 s	10–20 s	20–30 s	>30 s	Minimum value	Maximum value	Average value
Nonpeak hours	Walking-bicycle	60.1	20.2	8.6	11.1	0	185	19.7
	Walking-bus	65.5	20.6	4.7	9.2	0	174	16.5
	Walking-car	72.8	11.8	6.5	8.9	0	130	14.3
Peak hours	Walking-bicycle	53.5	22.3	11.3	12.9	0	233	21.6
	Walking-bus	43.8	10.1	14.5	31.6	0	744	57.4
	Walking-car	37.5	12.5	12.5	37.5	2	675	67.8

7.5 Sensitivity Analysis of Trip Chain Information Recognition Based on Simulation Data

Because large-scale field mobile sensor data require considerable workforce and time costs, this section uses the "traffic communication" integrated simulation platform to generate simulated mobile sensor data. By setting the same route and traffic environment as those used for the field data, the simulation data and the field data can be made consistent. To verify the reliability, the previous chapters' algorithm is used to identify the traffic mode based on the simulation data.

7.5.1 Sensitivity Analysis of Travel Mode

In this part, through the comparative analysis of the algorithm recognition results and the travel information loaded in the simulation platform, the evaluation and error analysis of the traffic mode recognition effect is conducted in terms of the recognition accuracy.

1. Selection and Influence of the Algorithm

Table 7.4 shows the statistical results of traffic mode identification under different algorithms. The algorithms used include a neural network, support vector machine, Bayesian network, random forest, four separate machine learning algorithms, and four integrated models that combine an algorithm with GIS map matching. Similar to the results when using data collected in the field, when only a single machine learning algorithm is used for traffic mode recognition, the walking, bicycle, and automotive vehicle recognition accuracy is greater than 90%. In contrast, the bus and car recognition results are not ideal, ranging from 60 to 75%. The main reason is that the bus and car's velocity characteristics are similar and difficult to distinguish. After the map matching algorithm is used, the recognition accuracy of the two traffic modes is significantly improved, reaching more than 90%. Compared with that for the real data, the recognition accuracy of the simulation data is 3–4%

Table 7.4 Statistical results of traffic mode identification under different algorithms

Traffic mode	Traffic mode recognition accuracy (%)							
	Machine learning algorithm				Algorithm combined with GIS map			
	Neural network	SVM	RF	BNs	Neural network	SVM	RF	BNs
Walking	98.7	98.7	99.1	98.2	98.7	98.7	99.1	98.2
Bicycle	93.4	95.6	92.5	94.2	93.4	95.6	92.5	94.2
Automotive vehicle	94.5	96.7	93.2	92.2	94.5	96.7	93.2	92.2
Bus	72.7	69.9	67.5	66.4	93.9	96.9	98.6	96.7
Car	67	69	71	67	96.5	98.3	94.7	98.3
Average	85.26	85.98	84.66	83.6	95.4	97.24	95.62	95.92

higher on average, mainly because the simulation test environment is less affected by external interference. Moreover, the map matching algorithm can improve traffic mode recognition.

2. Setting and Influence of the Simulation Data Sampling Frequency

On the basis of the above research results, the paper uses the SVM and map-matching algorithm to analyze the influence of the sampling frequency of simulation data. The results of traffic mode recognition under different data sampling rates are shown in Fig. 7.5. Similar to the data recognition results of field sampling, when the sampling frequency is high (1–5 s), the recognition accuracy of the four traffic modes is greater than 80%, especially for the high sampling frequency of 1–3 s, and the recognition

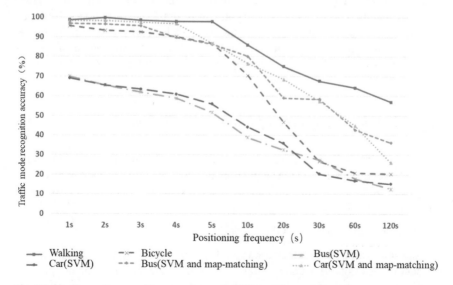

Fig. 7.5 Traffic mode recognition accuracy under different data sampling rates

Table 7.5 Traffic mode recognition accuracy under different traffic conditions

Traffic mode	Nonpeak hours (SVM)	Peak hours (SVM)	Nonpeak hours (SVM and map matching)	Peak hours (SVM and map matching)
Walking	98.7	97.3	98.7	97.3
Bicycle	95.6	92.6	95.6	92.6
Bus	69.9	64.3	96.9	92.7
Car	69	65.6	98.3	93.5

accuracy of traffic modes exceeds 90%. When the sampling interval is more than 20 s (more than 70%), the traffic mode identification accuracy decreases rapidly. When the sampling rate is more than 120 s, with the exception of the walking mode (60%), the identification accuracy is approximately 30%. The reason is that when the position sampling frequency is low, many features between two recorded points are not registered; in particular, bus stop information has a high probability of being missed. Therefore, in the practical application of GPS survey technology, considering the battery service time and data calculation cost, a sampling frequency of 5 s is recommended, and the maximum sampling frequency should not exceed 20 s.

3. Simulation of Traffic State and Impact

The real traffic flow in peak and nonpeak hours is simulated and analyzed using the SVM and map—matching algorithm. The statistical results are shown in Table 7.5. The recognition accuracy of the walking and bicycle traffic modes exceeds 90% in the peak and nonpeak periods due to the weak influence of traffic flow on these modes. When using only the SVM algorithm, the bus and car recognition accuracy is approximately 69% in nonpeak hours and drops to 65% in peak hours. However, after using the map matching algorithm, the accuracy is improved to more than 90%.

Compared with the results obtained for the real data, the simulation data's recognition accuracy is 3–5% higher on average due to the simpler traffic environment.

7.5.2 Sensitivity Analysis of Mode Transfer Time

In this part, through the comparative analysis of the algorithm recognition results and the travel information in the simulation platform, the effect evaluation and error analysis of traffic mode recognition is conducted using the recognition error of the travel mode transfer moment.

1. Selection and Influence of the Algorithm

The statistical results of the simulation mobile phone sensor data travel mode transfer moment recognition error under different algorithms are shown in Table 7.6. Four algorithms combined with GIS map matching are used to construct comprehensive

Table 7.6 Statistical results of the simulation mobile phone sensor data travel mode transfer moment recognition error for different algorithms

Algorithm	Travel mode	Absolute error distribution (%)				Absolute error (s)		
		0–10 s	10–20 s	20–30 s	>30 s	Min	Max	Mean
Neural Network + map-matching	Walking-bicycle	78.1	5.8	1.6	14.5	0	52	17.4
	Walking-bus	73.3	10.0	5.2	13.7	0	340	26.0
	Walking-car	83.6	6.5	0.0	10.0	0	180	12.0
SVM + map-matching	Walking-bicycle	78.3	6.7	1.7	13.3	0	52	16.9
	Walking-bus	71.6	11.2	4.1	13.1	0	340	25.8
	Walking-car	83.1	5.9	0.0	11.0	0	181	12.5
RF + map-matching	Walking-bicycle	80.1	5.7	1.6	12.5	0	52	13.1
	Walking-bus	70.9	10.1	4.1	15.0	0	342	23.0
	Walking-car	83.1	5.8	0.5	10.7	0	179	15.1
BNs + map-matching	Walking-bicycle	80.5	4.8	1.6	13.1	0	52	15.6
	Walking-bus	71.9	11.3	4.1	12.6	0	340	23.8
	Walking-car	83.1	7.9	0.0	9.3	0	183	12.6

models, and the sampling interval is 1 s. The recognition accuracy of different algorithms for the transportation mode transfer point is generally good, and the proportion of recognition errors that are less than 20 for walking—bicycle, walking-bus and walking—car is more than 70%. Furthermore, the average error is less than 30 s. Compared with that of the traditional questionnaire survey, the accuracy was greatly improved. The four algorithms produce similar results, and the advantages and disadvantages of different algorithms are not obvious. The difference between the average transfer time error of the real data and simulation data is less than 10 s, and the difference is less than 5.5% when it is more than 30 s.

2.   Setting and Influence of the Simulation Data Sampling Frequency

The statistical results of the travel mode transfer moment recognition error under different sampling frequencies are shown in Table 7.7. The SVM and map matching algorithm is used. Clearly, the distribution characteristics of the error percentage change significantly under different sampling frequencies. When the sampling frequency is high (1–5 s), the proportion of travel mode transfer moment recognition errors that are within 10 s is 50–70%, and the proportion of errors greater than 30 s is less than 30%. As the sampling frequency decreases, the travel mode transfer moment recognition error increases significantly. When the sampling interval is more than 60 s, the proportion of errors that are less than 10 s decreases to less than 20%, while the proportion of errors that are greater than 30 s increases rapidly to more than 50%. The overall error trend shows that with decreasing sampling frequency, the data become sparser, and the travel mode transfer moment recognition error gradually increases.

Moreover, as the in-sampling frequency decreases, the travel mode transfer moment recognition average error increases gradually. When the sampling frequency is high, i.e., 1–4 s, the travel mode transfer moment recognition average error of walking-bicycle is larger than that of walking-bus and walking-car. When the sampling interval is greater than 5 s, the travel mode transfer moment recognition average error of walking-bus and walking—car exceeds the recognition error of walking-bicycle. The sensitivity of the sampling frequency for walking—bus and walking—car is greater than that of walking-bicycle. Similar to the results for the real data, when the position sampling interval is not more than 10 s, the average error is less than one minute. When the position sampling interval increases to 120 s, the travel mode transfer moment recognition average error is more than two minutes. That of walking-bicycle. Similar to the results for the real data, when the position sampling interval is not more than 10 s, the average error is less than one minute. When the position sampling interval increases to 120 s, the travel mode transfer moment recognition average error is more than two minutes.

3.   Simulation of the Traffic State and Impact

In this section, the support vector machine and map matching algorithm are used to analyze the simulation data's travel mode transfer moment recognition error in nonpeak and peak periods. The results are shown in Table 7.8. The travel mode transfer moment recognition error, especially that of walking-bus and walking-car,

Table 7.7 Statistical results of the travel mode transfer moment recognition error under different sampling frequencies

Algorithm	Travel mode	Absolute error distribution (%)				Absolute error (s)		
		0–10 s	10–20 s	20–30 s	>30 s	Min	Max	Mean
1 s	Walking-bicycle	78.3	6.7	1.7	13.3	0	52	19.9
	Walking-bus	71.6	0.0	4.1	24.3	0	340	14.8
	Walking-car	83.1	5.9	0.0	11.0	0	181	12.5
2 s	Walking-bicycle	78.3	5.0	3.3	13.3	0	53	21.4
	Walking-bus	74.7	0.0	2.7	22.6	0	162	17.2
	Walking-car	97.5	0.0	0.0	2.5	0	93	15.6
3 s	Walking-bicycle	66.7	15.0	5.0	13.3	0	74	25.2
	Walking-bus	74.3	0.3	2.0	23.3	0	165	21.7
	Walking-car	92.4	0.8	0.0	6.8	0	126	29.8
4 s	Walking-bicycle	66.7	11.7	3.3	18.3	0	185	33.3
	Walking-bus	72.6	0.7	2.4	24.3	0	164	33.1
	Walking-car	84.7	6.8	0.0	8.5	0	118	30.9
5 s	Walking-bicycle	65.0	8.3	10.0	16.7	0	56	35.0
	Walking-bus	68.2	5.7	1.7	24.3	0	188	44.6
	Walking-car	72.0	19.5	0.0	8.5	0	131	41.8
10 s	Walking-bicycle	61.7	25.0	0.0	13.3	0	51	48.0
	Walking-bus	57.4	36.1	1.4	5.1	0	132	51.3
	Walking-car	55.9	41.5	2.5	0.0	0	27	48.5
20 s	Walking-bicycle	40.0	8.3	21.7	30.0	0	61	59.0
	Walking-bus	44.9	14.5	12.2	28.4	0	153	63.7
	Walking-car	50.8	5.1	28.0	16.1	0	69	66.4
30 s	Walking-bicycle	30.0	21.7	11.7	36.7	1	225	63.2
	Walking-bus	32.1	9.1	17.9	40.9	0	252	85.9
	Walking-car	38.1	13.6	3.4	44.9	0	181	88.0
60 s	Walking-bicycle	16.7	18.3	13.3	51.7	2	285	66.8
	Walking-bus	19.1	7.1	10.5	63.3	0	592	92.7
	Walking-car	13.6	12.7	18.6	55.1	0	231	96.1
120 s	Walking-bicycle	21.7	1.7	3.3	73.3	7	646	152.3
	Walking-bus	5.1	22.0	4.4	68.5	1	1069	155.5
	Walking-car	9.3	24.2	8.5	58.0	1	459	134.4

is much higher under congested conditions. The average recognition error of walking-bus increases from 25.8 to 86.4 s, and the average recognition error of walking-car increases from 12.5 to 76.2 s. In terms of the absolute error distribution, the proportion of errors that are within 0–10 s is reduced by 20–50%, and the proportion of errors that are more than 30 s is increased by nearly 30%. The impact of traffic congestion on

Table 7.8 Travel mode transfer moment recognition error during nonpeak and peak periods

Traffic state	Traffic mode	Absolute error distribution (%)				Absolute error (s)		
		0–10 s	10–20 s	20–30 s	>30 s	Min	Max	Mean
Nonpeak hours	Walking-bicycle	78.3	6.7	1.7	13.3	0	52	16.9
	Walking-bus	71.6	11.2	4.1	13.1	0	340	25.8
	Walking-car	83.1	5.9	0.0	11.0	0	181	12.5
Peak hours	Walking-bicycle	70.5	8.6	6.4	14.5	0	368	29.4
	Walking-bus	49.1	4.5	0.0	46.4	0	862	86.4
	Walking-car	37.4	13.4	11.7	37.5	0	429	76.2

the walking-bicycle travel mode transfer moment recognition error is not substantial. As mentioned earlier, this is mainly because traffic congestion has little impact on walking and cycling.

7.6 Summary

Traffic survey technology based on GPS positioning, which has been the focus of research and field applications in traffic surveys in recent years, can automatically obtain accurate, dynamic and real-time individual travel trajectory data. Through the design and collection of field test data, this chapter evaluates, in detail, the effect and influence of different data mining algorithms (neural network, support vector machine, random forest, Bayesian network), sampling intervals (1–120 s), and traffic states (nonpeak hours and peak hours) on the practical application of traffic survey technology based on GPS. The research results include the following three aspects:

(1) The four algorithms mentioned can be used for travel chain information recognition, especially for walking, bicycles, and automotive vehicles. The accuracy of travel mode recognition is approximately 90%. However, for buses and cars, the accuracy is not ideal (approximately 60–70%). The proposed map matching algorithm based on bus stop GIS information can compensate for the single algorithm's shortcomings and improve the recognition accuracy of buses and cars to more than 90%.

(2) The data sampling frequency significantly affects the effectiveness of traffic survey technology based on GPS positioning. When the sampling frequency is high (1–5 s), the traffic mode recognition accuracy is more than 80%, and the travel mode transfer moment recognition error is mostly less than 20 s. As the data sampling frequency decreases, the accuracy decreases rapidly. When the sampling interval is more than 120 s, the traffic mode recognition accuracy is generally less than 30% (except that of walking, 60%). The average error of the walking-bicycle transfer moment recognition is close to 2 min, and that of walking-bus and walking-car is close to 3 min.

(3) Traffic congestion affects the application of GPS survey technology. When using the machine learning algorithms, the traffic mode recognition accuracy in the congested state is 10–20% lower than that in the unblocked state. The walking-bus and walking-car transfer moment recognition average errors are approximately 40 s, and the walking-bicycle errors are almost the same in the unblocked and congested states. This is mainly due to the limited impact of traffic congestion on walking and bicycling. In addition, GIS-based bus stop matching technology can significantly improve the recognition accuracy of bus and car modes under congested traffic conditions.

(4) The four machine learning algorithms can be used to recognize the simulation data's travel chain information, and the recognition results are similar to those for real sensor data. The recognition accuracy of walking, bicycling and motor vehicles exceeds 90%. However, the recognition of buses and cars is not ideal (60–75%). The recognition accuracy can be significantly improved to more than 90% by the GIS map matching algorithm.

The sampling frequency also has a considerable influence on the travel mode identification of simulation sensor data. When the sampling frequency is high (1–5 s), the recognition accuracy of the four traffic modes is greater than 80%, especially for the high sampling frequency of 1–3 s, and the recognition accuracy of the four traffic modes exceeds 90%. When the sampling interval is more than 20 s (recognition accuracy more than 70%), the traffic mode identification accuracy decreases rapidly. When the sampling interval is more than 120 s, the recognition accuracy of all modes except walking (60%) is approximately 30%.

Similar to the results for real sensor data, traffic congestion has a substantial influence on the recognition and extraction of simulation data. The recognition accuracy of pedestrian and bicycle traffic modes is more than 90% in peak and nonpeak times. In comparison, the bus and car recognition accuracy is approximately 69% when using the support vector machine only, and it drops to 65% during peak times. However, the map matching algorithm improves the accuracy to more than 90%. Compared with that of the real data, the recognition accuracy of the simulation data is 3%-5% higher. Under congested conditions, the walking—bus transfer moment recognition average error increases from 25.8 to 86.4, and that of walking-car increases from 12.5 to 76.2. In terms of the absolute error distribution, the proportion of errors within 0–10 s is reduced by 20–50%, and the ratio of errors more than 30 s is increased by 30%.

Chapter 8
Thinking About Application of Refined Travel Data in Traffic Planning

For a long time, the quantitative relationship between transportation and urban development and social economy has not been established, so empiricism has the upper hand in the discussion of many urban transportation development problems and countermeasures. This approach achieves a particular effect for experienced experts and managers in the stage of urban transportation problems, which are not incredibly complex. Nevertheless, with the development of urban transportation problems, it is urgent to use refined and accurate individual travel survey data and analysis results to support the development of a quantitative rational decision—making model. The product also promotes the transformation and upgrading of the transportation development decision—making mode from "empiricism" to "rationalism".

With the advent of big data, the mining and analysis of massive data will provide more valuable information to assist decision making and has become a hot research frontier in all walks of life both domestically and abroad. However, despite the popularity of big data research, we need to think deeply about the specific uses and functions of data. Bubbles and impetuosity will accomplish nothing. Therefore, we can not only remain in the joy of "soap bubbles" achieved by big data resources but also need to dig deep into data to produce practical value in applications.

This chapter focuses on the application of refined data on individual travel activities. Fine calibration and optimization are performed from the perspective of a theoretical model. The model includes three components: verification and calibration of the traditional four—step method to improve the accuracy, evaluation and optimization of the bus OD backstepping model, and empirical calibration of the new generation of traffic demand analysis models based on activities. Because traditional travel surveys cannot provide the detailed travel chain data needed for activity model calibration, they have not been widely used to conduct traffic demand analysis through empirical calibration activity models. The refined data of individual travel activities can support the construction of an activity—based traffic demand model by collecting travel chain data.

© Tongji University Press 2022
F. Yang and Z. Yao, *Travel Behavior Characteristics Analysis Technology
Based on Mobile Phone Location Data*, https://doi.org/10.1007/978-981-16-8008-3_8

Section 8.1 analyzes the precision improvement for the traditional four—step method by means of refined data, including the application of data precision improvement from traffic district subdivision, traffic mode identification and evaluation of public transport OD backstepping model, real travel path checking and auxiliary traffic allocation. Section 8.2 analyzes the application of refined data in optimizing the layout of bus stops and networks. Section 8.3 demonstrates the process of the activity—based traffic demand analysis model. It discusses the application of refined data in constructing the traffic model based on refined individual activities. Section 8.4 analyzes the application of refined individual travel data to extract individual activity rules in urban planning, building site selection, hot spot safety detection and other aspects.

8.1 Optimizing the Traditional Four-Step Method

The four—step method of urban transportation is taken as the research object. Under the same conditions, this paper conducts a test to compare the accuracy of the traditional resident questionnaire survey model and the smartphone app survey under different refinement degrees of traffic district division. This approach can provide an empirical basis from a quantitative perspective to assess the accuracy improvement effect of refined travel data on the four—step method. The research results are expected to provide a reference for revising existing and building new urban traffic models.

(1) Traffic Districts are Subdivided Into "Traffic Subdistricts" to Improve the Accuracy of the Traditional Four—Step Method.

The more detailed and smaller the traffic area is, the more accurate and complete the expression of residents' travel activities. However, in the actual modeling application, to facilitate the household sampling survey, areas are often divided according to administrative divisions or community neighborhood committees, with a wide range of areas and a small number of traffic districts. This method may result in some internal travel demand not being expressed. A large amount of travel information is ignored, resulting in incomplete travel data, which makes it challenging to satisfy the accuracy requirements of the four—step method. Most large—and medium—sized cities in China can be divided into several hundred traffic districts. Each traffic district area consists of 1–3 km^2 in the dense downtown area, and the scale of the peripheral area may be larger. At present, there are more than 1000 traffic districts in the traffic models of a few megacities in China and approximately 2000 in Shanghai.

The GPS positioning accuracy of a mobile phone is usually 5–10 m and differs according to the GPS chip's quality. The smallest plot in an urban center intensive development area is generally 300 m * 300 m = 0.09 km^2, close to 0.1 km^2. The GPS positioning error is far less than this, so positioning error has little influence on the traffic area coding of travel destinations. The refined individual survey based on the mobile app can support traffic area division into smaller dimensions called "traffic subareas". The more refined division can result in more traffic districts; therefore, more refined individual travel data can be obtained. The GPS coordinates of the OD endpoints can be extracted to overcome the shortcomings of the traditional questionnaire survey and to obtain the travel demand that cannot be expressed in the original traffic area, which will greatly improve the accuracy of the four—step method.

As shown in Fig. 8.1, different expressions apply after the traditional traffic district is divided and refined into traffic subdistricts. Figure 8.1a shows the OD expectation line between two conventional traffic districts. The area of traditional traffic districts is large, and there is a broad expectation line between the two communities. Figure 8.1b shows the OD expectation line for each traffic subarea after the traffic subarea is subdivided into multiple traffic subareas. These results more precisely express the traffic volume between two traffic subareas. Figure 8.1c shows the OD expectation line of each traffic subinterval in the original traffic district, which is accurately represented.

(2) Traffic Mode Recognition and Evaluation of the Public Transport OD Backstepping Model

The public transport sharing rate is an important index to measure public transport development and the rationality of the urban traffic structure. Refined individual travel data can accurately identify different transportation modes, determine transportation modes for specific sample sizes, screen out bus travel data, and calculate the bus share rate.

There are two main ways to obtain bus OD. One is the car—following survey, but this method is not universal due to the shortcomings of heavy workload and complicated organization. The other is based on the bus card data OD backstepping model. The current research results show that the backstepping accuracy is not ideal, approximately 30–40%. Moreover, some individuals do not use bus IC cards, making it difficult to accurately assess the OD of bus travel.

The mobile app can collect complete bus travel trajectory data with high accuracy. By mining and analyzing mobile phone sensor data, we can accurately identify the complete trajectory of individual bus travel and obtain the complete OD data between individual bus stops using the bus mode. Tracking is also possible for individuals who do not use the bus IC card. Comparing the refined benchmark data with the results of various bus OD backstepping models can help to select an appropriate backstepping model for different real situations and then calculate the bus share rate. In addition, accurate data from small samples can help to reduce the sampling rate. Initially, more sampling rate data analysis is needed. With the improvement of survey data quality, less sampling survey data may be sufficient to accurately represent the real situation of public transport passenger flow.

(3) Real Travel Path Verification Based on Refined Individual Travel Data

The traditional path assignment method based on the path impedance mode has some defects. The traffic assignment of the four—step method is based on travel time or cost as an impedance function, which is usually quite different from the

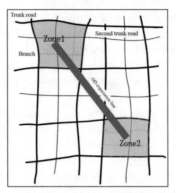

(a) OD expectation line between traditional traffic districts

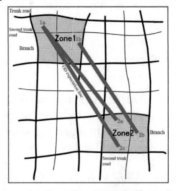

(b) External expectation line of each subdistrict in the traffic district

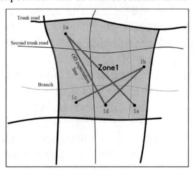

(c) OD expectation line of the traffic subarea

Fig. 8.1 Detailed expression of travel when traffic areas are subdivided into traffic subareas

actual travel route choice. For example, the travel route may temporarily change due to terrain (mountain), real—time congestion and other reasons, which will cause the theoretical shortest path to not be the real travel path and lead to deviation in the traffic assignment results. Refined survey technology can be used to accurately extract the individual travel chain, and the adopted traffic mode, travel time and other information can be used to modify the impedance function and check the real travel path. Moreover, the effect of the number of individual travel refinement data on the traffic model accuracy can be assessed from the perspective of a quantitative empirical comparison.

The number of trips can also be accurately identified by checking the route, especially for short—term and short—distance trips. In traditional resident travel surveys, short—term and short—distance travel surveys of residents often have low survey accuracy. The refined individual travel data can better identify the short—term and short—distance travel of residents. The verification method adopts the comparative research method, taking a city as the background and using the household travel questionnaire survey and mobile data collection app to simultaneously obtain data from two sources. The data include travel data D1 from the traditional residents' travel questionnaire survey and mobile sensor data D2 from the mobile data collection app.

The accuracy of the individual travel data of D1 and D2 is compared, accurate statistics of the travel times are obtained, and the change effect of data quality compared with the traditional questionnaire survey is comprehensively evaluated. Furthermore, the two data sources are imported into the urban traffic model to obtain the respective flow allocation results V1 and V2. Assuming the other conditions are the same, the accurate collection of urban traffic flow vs. the verification line is used to assess the accuracy of different travel time models under the two data sources.

8.2 Optimizing the Layout of Bus Stations and Network

Prioritizing public transport development is the fundamental means to alleviate traffic congestion in China's large cities. There are many unreasonable network layouts and station distances in the public transportation network in real life. Optimization of the public transportation network can facilitate travel. In the construction of city transit, refined individual data can be used to optimize the bus stops and routes, guide the bus operation, ensure successful transfer connection, and set reasonable station spacing and departure spacing.

At present, the distances between public transport stations in many cities are 300 and 500 m, and the coverage rate is high. However, this does not mean that residents can easily travel (for example, if there is no bus line nearby). Therefore, it is essential to accurately understand individual travel paths and address public transport travel difficulties to improve residents' public transport travel willingness and public transport share rate. Refined data can be used to accurately track and obtain the process and space—time distance from the public transport station to the destination. It is helpful to fully understand the source and destination of passenger flow at the station, master the main passenger flow absorption and departure points along the public transport line, and reveal the last kilometer's characteristics. It is of great significance to guide the connection between rail transit and buses and optimize the existing bus network. Figure 8.2 shows the route of the original bus 83, the starting and ending points of residents' travel, and the bus stop location distribution. The yellow line represents the direction of the original bus line. The red origin indicates the bus stop, and the yellow origin represents the starting and ending points of residents' travel. Detailed data analysis shows that most passengers who get off at Qingyang Avenue South Station must walk a long distance to reach their destination, which is inconvenient. Figure 8.3 shows the location distribution of bus stops and residents' starting and ending points after optimizing the bus route. The blue line represents the optimized bus route. After optimization of the bus route, the travel distance of residents is significantly shortened, and the path is more in line with the travel needs of residents.

The source and destination of subway passengers can be obtained using the handover sequence of communication base station cells to track the location change information of subway passengers, combined with the base station's geographical location. Furthermore, this information can be used to analyze rail transit stations' attraction scope and to determine the passenger flow number, which can guide rail transit and bus transfer. Figure 8.4 shows the station location distribution on the metro line and the starting and ending points of residents' travel. The red line represents the rail transit line, the red dot is the rail transit station, the yellow dots are the starting

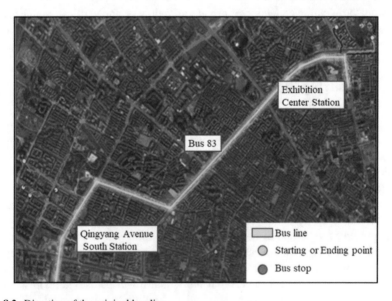

Fig. 8.2 Direction of the original bus line

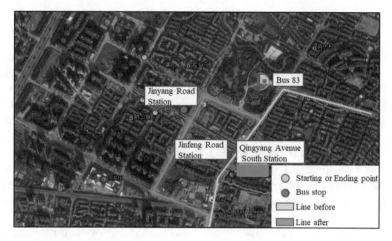

Fig. 8.3   Optimized bus route

and ending points of residents' travel, and the circular area surrounded by the white line is the passenger flow attraction scope of the station. The attraction scope of each rail transit station can be obtained by analyzing the refined data.

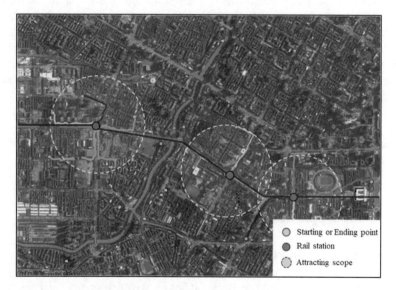

Fig. 8.4   Passenger flow attraction scope of the rail station

8.3 Constructing Activity Based Traffic Demand Model

In recent years, the field of transportation has explored new models to replace the four—step method. Activity—based traffic demand analysis has become a research focus. However, suppose the development of this method wants to achieve a leap from theory to practice. In that case, it must be supported by sufficient precise and accurate individual travel activity characteristic data, including specific departure time, arrival time, activity type, and residence time. However, in the actual travel survey, many data biases exist, and short—distance travel is missing. Moreover, the respondents ignore several travel activity characteristics. The traditional questionnaire survey method cannot be used to obtain precise and accurate data on individual travel activity characteristics. The data quality is far from meeting the requirements of the traffic demand model.

Refined data can be used to accurately extract individual travel information about various activities to obtain complete travel chain information based on activities. On the basis of the microscopic simulation technology in the activity model, individual trip chain and frequency in one day can be predicted. Traffic mode prediction, stopping point prediction, starting and ending time and duration prediction in the travel chain can be achieved by discrete choice models, intermediate stop location models, Monte Carlo simulation methods and risk duration models, respectively. Then, this information can be integrated into the OD matrix to complete the traffic assignment.

The mobile app's survey data can also be used to check and evaluate the prediction accuracy based on the activity model. The main ideas are as follows. First, the individual mobile base station conversion sequence $(L_1 - L_2 - \cdots - L_n)$ is obtained from the call duration and base station information in the mobile data. Second, the real base station position trajectory $(L_1 - L_2 - \cdots - L_n)$ is obtained by setting a time window. Then, the base station's location trajectory is transformed into the real travel location trajectory of the individual using the individual call probability P_i combined with the linear equation. Finally, this information is matched with the prediction results based on the activity model through the sequence alignment method to assess the accuracy of the activity model's prediction results.

8.4 Other Applications

The new mobile data are expected to achieve fine extraction of travel information to better reflect the origin of individual space—time activities. This process is of great significance for traffic demand analysis and has great expansion potential and application value for urban planning, commerce, tourism, and public security antiterrorism management. For example, the data can be used for the development and construction time sequence of a new area, the analysis of job housing balance, the relocation and construction of major urban land facilities, the analysis of customer activity behavior in the process of a shopping mall or commercial exhibition, and the determination of the stay time and visiting order of tourist attractions.

(1) Development and Construction Time Sequence of a New Area Guided by Refined Individual Travel Data

The refined individual travel data provide important support for describing the spatial—temporal law of residents' travel activities. The data are a useful supplement to traditional research methods of urban spatial structure and can provide an important decision—making basis for urban construction and development. Land use and development projects in the process of development and construction of new areas in large cities are gradually promoted according to the time sequence. It takes a long time for commercial projects to become popular. Because of the uncertainty of traffic demand in a new area, it is challenging to allocate traffic services, for example, determining bus lines, departure frequency, and the size of the public bicycle configuration. Moreover, the dynamic travel demand in the new area's development process is difficult to obtain by means of the traditional resident travel survey method. In contrast, mobile sensor survey technology collects information such as the individual movement trajectory and the number of people in the base station community. This information is helpful to understand the characteristics of all—weather travel activities within the new district's scope and the areas related to the new district to flexibly configure transportation support services and avoid the waste of idle facilities.

(2) Site Selection of Large Buildings Based on Refined Individual Travel Data

Important business districts, hospitals and transportation hubs of a city have a significant impact on residents' travel. They are the distribution centers of many people in the city and are important nodes of urban transportation. The selection of these sites selections has a significant impact on the operation of urban functions. Mobile phone data are used to analyze the spatial—temporal law of residents' travel. On the one hand, site selection can determine the urban function of important urban facilities and the scope of residents' gathering and scattering attraction. When a location needs to be adjusted, a new location can be quickly and scientifically selected to improve residents' spatial and temporal accessibility. On the other hand, according to the number of mobile phones and individual mobile phone sensor data in the facility area, the residents' travel attraction level, time and selected traffic can be calculated to establish the layout, connection and scheduling scheme of transportation facilities. Furthermore, according to the number of mobile phones and individual mobile phone sensor data in the facility area, we can calculate the attracted travel volume, time and selected travel mode of residents to formulate the layout, connection and scheduling scheme of traffic facilities. For example, when a large hospital in Chongqing is considering relocation due to the need to increase usable land, the planning and Research Department of Chongqing can determine the source and flow direction of patients and doctors and the traffic mode structure based on mobile signaling data. These data provide a decision—making reference for the hospital's relocation site selection, timely adjustment of the transportation facilities, and minimization of the impact on the existing doctors and patients. Therefore, this approach realizes the joint scientific decision making of urban planning and transportation planning.

(3)   Refined Individual Travel Data Monitoring of Public Security in Hot Spots

Individual mobile phone big data can reflect high concentration areas of urban pedestrian activity and provide early warning support for urban public safety management, antiterrorism and stability maintenance. This area is an innovative application of mobile Internet technology in national stability maintenance and antiterrorism, specializing in national security. When a mobile phone is turned on, it maintains periodic contact with the communication network base station. Through the load characteristics of the communication activities and the number of registered mobile phones in the base station, the number of mobile phones in a crowd and the regional population density in the base station's coverage area can be determined to a certain extent. A monitoring system of people flow density in urban hot spots can then be established based on mobile phone data. Once a certain density is reached, emergency plans at all levels should be taken to various dangerous situations, thereby improving the level of urban public safety management. Shopping malls, squares, large—scale transportation hubs and other areas with concentrated pedestrian flows may become targets of terrorist attacks. The establishment of a monitoring system can facilitate timely deployment of the police force to strengthen the security in areas with high pedestrian flow densities and to handle unexpected mass incidents in high—density activity areas.

Chapter 9
Outlook

9.1 Technical Efficiency and Universal Upgrading

On the basis of extensive empirical data, this paper deepens the method system of individual travel chain information extraction based on mobile phone sensor data from three aspects, namely, improving the integrity of data content, the universality of the method and the effectiveness of the practical application.

(1) Traffic planning schemes are usually based on the accurate analysis and prediction of current traffic problems and traffic demand. The decision—making process is restricted and influenced by many fields and factors, such as transportation, economy and land use. In terms of travel information, it is often necessary to obtain individual detailed travel time, travel distance, travel OD, transportation mode, activity purpose, and other elements through household questionnaires, which is a cumbersome and complex process. The previous chapters of this book focus on the process and method of individual travel chain information extraction based on mobile sensors. In contrast to the traditional questionnaire survey method, this paper proposes a new approach to individual travel chain information extraction. However, a difficult problem remains to be solved: the identification of travel purpose. Unlike many European and American countries and cities, due to historical development, urban scale and other reasons, the composite degree of urban building development in Europe and America is much simpler than that in China. Therefore, based on the nature of land use and mobile phone sensor individual travel location-tracking information, foreign countries can better understand each trip's purpose. However, for China and other regions with large populations and high—density and complex construction development, even if accurate individual travel OD information is obtained, travel purpose is still difficult to infer. For example, when customers enter a complex building, their travel purpose may be dining, shopping, or meeting friends. Therefore, in subsequent technology improvements,

© Tongji University Press 2022
F. Yang and Z. Yao, *Travel Behavior Characteristics Analysis Technology Based on Mobile Phone Location Data*, https://doi.org/10.1007/978-981-16-8008-3_9

the research on travel purpose identification should be supplemented and deepened to develop a complete set of individual travel chain information collection processes and methods based on mobile phone sensor data.

(2)   As mentioned above, this book's empirical data are mainly in the form of sample data collection and analysis from a small part of Chengdu. There are inevitably some deficiencies in the universality of the method, which cannot cover all areas (such as mountain cities) and all travel mode combinations. Furthermore, there is a lack of evaluation of the technology's application effect in the city. Before applying the technology, an extensive range of tests should be conducted on existing technical processes and methods. Alternatively, the proposed method should be combined with urban resident travel surveys, public transport planning, rail passenger flow forecasts and other practical projects to evaluate the effect of technology application and optimize the method.

(3)   The personal travel chain information collection technology based on the mobile phone sensor proposed in this book has a large amount of practical application data, and the maximum positioning interval can even achieve 1 s data collection. For a city survey, the amount of data is large. Thus, data processing and management must be explored with respect to cloud computing, distributed storage and other aspects. In particular, the selection of model parameters may also need to be based on the data acquisition time (peak, flat), positioning frequency, algorithm training depth, etc.

9.2   Multiple Heterogeneous Data Integrating

Transportation planning and decision making is a process of extensive and systematic research. With the coming of the big data and mobile Internet era, the application of big data to support urban traffic solutions has become a recognized planning concept and technology development trend. However, technology popularization and applications are still in the primary stage. The application objects are mostly traffic status analysis and macro decision analysis support. The main reason is that it is difficult to collect micro and fine travel information, especially complete individual travel chain information. This book introduces individual travel chain information collection technology based on mobile phone sensor data to provide preliminary technical exploration and development prospects for refined traffic big data collection. However, some shortcomings and defects remain.

Single data sets have difficulty representing full travel chain information on an individual travel. For example, for a trip completed by the combination of walking—bus—subway—walking, the mobile sensor can obtain accurate trajectory data of walking and bus travel. However, such data cannot be obtained for the subway due to the shielding of the mobile GPS sensor signal. The subway travel section, especially transfer information in the subway, cannot be obtained. Fortunately, mobile signaling data can compensate for this issue. Communication operators along the subway can be equipped with special mobile communication base stations that can interact with subway passengers regularly and irregularly to indirectly obtain the user's travel trajectory information. For another example, subway card data can help solve the bus card data problem to infer the location of getting off; wireless Wi-Fi data and acceleration data can help infer indoor travel trajectory and travel purpose.

Therefore, the fusion of multiple types of traffic data should be studied. Data fusion can even involve multifield and multidisciplinary data fusion. Simultaneously, exploring different disciplines to form common values and generate new cooperation modes and mechanisms is also a major focus in developing big transportation data.

9.3 Traffic Planning Theories and Models Upgrading

The concept of big traffic data has been around for many years. However, in the field of traffic planning, the practical application of traffic big data (including floating car data, mobile phone location data, bus card data, video/coil detection data) still lies in the analysis of traffic status, such as medium and macro travel rules. Major institutions have drawn many "beautiful" dynamic and static renderings, but the traditional planning model is still the main means of traffic demand forecasting. This model does not essentially improve the effect of traffic planning. The "aesthetic fatigue" of big data will come soon. At this stage, exploring new theories and traffic planning methods based on refined traffic survey big data will be a major focus.

This book discusses technical methods for the application of mobile phone sensor data in fine traffic travel chain information collection. This paper discusses the application of refined travel chain information for the improvement of traditional planning models (for example, traffic district subdivision, travel path selection) and traffic planning practice (bus line optimization, residents' "last mile" optimization). As a good example, we also advocate the further study of traffic planning models based on refined traffic survey data and traffic big data and exploration of a new traffic planning model based on activity.